Purposeful Birdwatching

A little valley on the edge of the New Forest: swallows slinking over the flowery meadow, cattle in the pasture, a buzzard rising from a fence post, the high forest beyond. It could almost equally be mid-Staffordshire, or the Welsh borders. Ordinary, but extraordinary; a beautiful bit of England.

PURPOSEFUL BIRDWATCHING

GETTING TO KNOW BIRDS BETTER

ROB HUME

with illustrations by the author

PELAGIC PUBLISHING

First published in 2024 by
Pelagic Publishing
20–22 Wenlock Road
London N1 7GU, UK

www.pelagicpublishing.com

https://doi.org/10.53061/TQVW2957

A CIP record for this book is available from the British Library

ISBN 978-1-78427-468-9 Paperback
ISBN 978-1-78427-469-6 ePub
ISBN 978-1-78427-470-2 PDF

Typeset in Scala by BBR Design

This book is dedicated to Carolyne, Noah and Charlotte, busy building their own memories, and to my wife Marcella, always ready with so much encouragement and support.

I particularly offer my sincere thanks to the late Roger Lovegrove and to Trevor Gunton, who, against all expectations, helped me turn a hobby into a career; and to Nick Hammond for later getting me into writing and editing.

PUT THIS BOOK DOWN for a minute and go to the back door – or a window, a balcony, whatever you've got – and take a look outside.

What did you see? Most likely a pigeon or two, maybe a black 'crow' or a passing magpie, perhaps a blackbird on the lawn or a nearby aerial, and probably something smaller, flitting about on the edge of a flowerbed or half-hidden in the foliage of a tree. Depending on your birdwatching abilities (which may or may not relate to the length of time you have been doing this) you may identify everything, or most, or just a few of the birds. In this book, I hope my experience over more than 50 years can help or point you towards new ideas, ring a few bells, bring a few nods of agreement, even make you shake your head now and then, depending on where you are on your route to expertise. It doesn't matter: this book is for anyone interested in birds, whether starting out, at the pretty-good-at-it stage, or already what most people would regard as expert.

Being expert, to the average onlooker, means being able to tell one bird from another. The better you get, expertise tends to be increasingly associated with identifying birds with a briefer view, or at a greater distance. But telling what a bird is, that is just the beginning. Many questions arise, beyond simple identification, if you have an enquiring kind of mind: not just *what*, but how and why. Dylan Thomas wrote that his Christmas gifts included a book that told you everything you wanted to know about wasps, except *why*. If nothing else, birdwatching makes you think.

Keep an open mind, keep asking questions. Question the established 'perceived wisdom', sometimes. Most of all, appreciate what you see and enjoy it, whatever it might be. An almost childlike fascination with the world and all the remarkable, beautiful things within it will serve you well. Who knows where you will go with it?

CONTENTS

PREFACE

This is unlike the average bird book (I've written a few average bird books over the years, so I ought to know). You can read it cover to cover, or just dip into long or short items at random... please *just browse* if that is your preference.

The book is a mix of personal memoir and gentle help and advice for those who welcome it. Extracts, more or less verbatim, from my old notebooks are shown in a different font for clarity. They are varied, deliberately, to show the kind of thing that an ordinary birdwatcher *can* see with a bit of effort, and to reveal especially to younger readers what *could* have been seen just 20, 30 or 50 years ago: many things have been lost, while others have been gained. Predictions of extra southern species appearing because of climate change have mostly not yet transpired, so the losses generally outweigh the gains.

There's a long section focusing on birds at a local 'patch' – always a required element of any birdwatcher's toolkit and a sound basis for further study, if you can find a good one. Growing up through school days and student vacations, I was lucky to live near a reservoir with fascinating surroundings – Chasewater in Staffordshire – and later I found new places close to hand such as Blackpill in Swansea Bay and Blashford Lakes in Hampshire. Then there are birds seen on holiday and birds seen on days out especially to visit exciting new habitats and new places; or in some cases on travels specifically to see rarities (the real meaning of twitching, a term now rendered meaningless by its persistent misuse in the media).

What you will find here is not at all intended to say that you *should* do this, or that, or *must* watch birds in the way that I imply. It seems reasonable to me that most people would like to identify what they see, but it is also suggested by others that even this is not really necessary. You can, if you like, just ask someone else and not bother with all those heavy identification books. There are suggestions, a bit of advice, but you will find your own way if you have not already done so.

There are, I know, too many bits about rarer birds or obscure ones, and especially gulls. I hope this adds interest for some readers – but, equally, those bits might be skipped by others (for the time being – you could get to like gulls eventually, honest; I love them).

If this encourages more people to go out and enjoy birds, to help understand them and how our birdlife is changing – and most especially to *care* for their future and the future of the natural world of which we are just one part – that will be a good result. Someone needs to care, and it would be good if it were you. There is, sadly, a realisation among the older conservationists around that we have been banging on about all of this for decades. 'Save the rainforest' has been a slogan all my life, yet world deforestation continues at a faster pace than ever before. Surely, now, things must start to improve?

The book inevitably shows what birdwatching used to be like: using simple books, basic and often inefficient communication with others. Not so long ago, too, the countryside offered fields full of finch, sparrow and bunting flocks. Fields where golden plovers gathered every winter. Hedgerows with willow tits and yellowhammers and tree sparrows, where cuckoos called each summer. These things have gone from many places, even if the fields and hedgerows remain. Younger readers may not realise; older ones might cry.

Everyone can be an expert now. Look it up on Wikipedia. Use the app on your smartphone. Listen for the alerts on your mobile. That's brilliant, it really is: I look things up all the time. No-one wants to restrict information. Yet somehow when I grew into birdwatching, with far less information readily to hand, it stimulated a rather different approach – and I can't help feeling that in some ways it was good for me: not *better*, necessarily, certainly not giving me more knowledge, but helping to develop my own skills. It is too easy now to miss the 'growing up' process, learning as you go, and the loss of that might not be so useful if you want to learn and absorb. But if you would rather go direct to advanced level, why not? At the same time, many 'experienced' people, who have been watching birds for decades, never get to A-level: they still call a house martin a swift. Or get halfway but find some things difficult. Years of *interest* do not necessarily equate to *expertise*. But it doesn't much matter. No-one is being judged.

When I was at school in the 1960s, we weren't entirely devoid of information. My school library (thanks to the biology teacher) did have a copy of the 1930s five-volume 'Witherby' *Handbook of British Birds*, which remains a classic, brilliant and ground-breaking. The library even began to collect the volumes of Bannerman's *Birds of the British Isles*, which was a more discursive, heavyweight epic, but didn't quite achieve the same reputation although it had lovely, highly individual, paintings by George Lodge. But it was an adventure of discovery, one way or another, that would not quite happen now.

This book is intended to be a bit of fun as well as something that might help encourage, stimulate or improve an interest in birds. Birdwatchers generally need a little injection of fun – and many times I have had days out with friends that were great fun as well as great for birds. The book is, too, a bit of a reaction against media mis-portrayals of birdwatchers. We are not all twitchers, not all watching through out-of-alignment binoculars that create a figure-of-eight image, not all talking about getting a sighting (instead of seeing) or seeing birds flighting (not just flying). We don't go on about 'spotting'. Often 'our' language is imposed upon us. In the pages that follow I hope to encourage people to take an interest, to try to understand, and most of all to *appreciate* birds, which (like most things) are truly *remarkable*. Scarcely believable, when you really think about how they live, how they work, their speed of movement through the landscape relative to their size. They deserve a lot of deep thought and observation. Birds, in so many ways, are magical, like the rest of the natural world; like us.

A six-year period in Swansea opened up a whole new world of birds, epitomised by this ring-billed gull (right), which appeared amidst two or three thousand common gulls.

WHAT IS BIRDWATCHING, ANYWAY?

For more than 50 years, although I swear I am no older than 34, my reason for looking at birds has not changed: I like them. For some reason, I respond to them. How it all began, long ago, when I was five, I'm not sure. I listened to Romany on the radio, pretended to be David Attenborough, even then. My parents came from very different backgrounds: one from farming families on reclaimed land in Essex, who, after surviving the war, moved to find a job and start a family in the industrial Midlands; the other from a Staffordshire coal-mining area, in which the miners would often find a connection with the outside world through racing pigeons, or keeping caged finches (brown linnets, green linnets, seven-coloured linnets). Both had a good reason to try to get out into the countryside, and I benefited from being taken out and about into beautiful rural Staffordshire as often as possible. A bunch of us kids had a brief interest in birds' eggs but none of the others pursued things further, just me. The eggs thing soon stopped.

From time to time there has been a greater emphasis on this or that, a few years of chasing rarities (twitching, in a rather mild way mostly, but with plenty of effort and excitement along the way), a few migration holidays, a bit of survey work or whatever, thousands of hours looking at gull roosts, some holidays around the world, even leading tours to exotic places – but essentially, I have always looked at birds from an aesthetic point of view and at best with some sort of emotional response. I watch them like a frustrated artist really (painting is no relaxation for me, it's too difficult), but it has served me pretty well, all things considered.

Yet a famous birdwatcher – Ian (D. I. M.) Wallace – whom I knew in a small way and spoke to now and then – called me not so long before he died in 2021 and said I might be in a good position to promote what he called 'purposeful birdwatching'. He wanted to encourage people to make something a little more useful and lasting out of an otherwise ephemeral sort of hobby. He called it our

A dozen or so different eggs in a couple of little biscuit tins was about the extent of my collection when I was maybe eight or nine: most were 'swaps' or given to me, and my interest may only have lasted one or two seasons.

'hobby–science'. I also recall Ken Smith, expert RSPB researcher into all kinds of things, perhaps best known for his woodpecker studies, saying that anyone wishing to go further should just choose a species, then study it.

I haven't really, mainly because I never thought I was good enough. But I never trained as a ringer either, so colour-ringing, such an invaluable tool in population studies, was out of my reach. I used to find birds' nests – all kinds of exciting things – but even the joy of the gloriously sky-blue, inky-black-spotted eggs of a song thrush seems to be a distant memory now. So, not being a great nester reduced options for a close species study; as, perhaps, did available time. But all the same, putting a bit of purpose into your hobby – not simply paying something back, but adding a bit to our cumulative knowledge – seems to be worth encouraging. A simple way is to study a place, if not a species: just to watch, regularly and repeatedly. Even just counting things helps. Over the years, such basic data builds up a picture.

What is it that birdwatchers do all day? Why do they do it? Non-birdwatchers could be forgiven for wondering. Well, as you know, it can be one of many things. For many it is just a way to get out and about in the countryside: like fishing, it is a pursuit that brings you into close contact with other things, and those other things can be as much a part of the enjoyment as the birds themselves. Why bother, then, to fish? Or watch birds? Many people don't: enjoying the countryside is enough. Or they may climb mountains instead.

But birdwatching, and its bewildering array of associated subjects, increases your understanding of the world out there, natural and unnatural – like an appreciation of the physical geography and geology of a place, local culture and more distant history that influence buildings and land use, anything that helps you to appreciate why places look the way they do. It adds purpose to your time outdoors. And you meet new people, make friends, see great places, maybe you travel. It is a hobby, a science. It's good for your health. And the birds are wonderful.

Birdwatching expands your mind.

Very often it starts with some sort of influence from family or friends, maybe even a mildly competitive element quite early on. There used to be letters and postcards. 'This week I've seen up to six long-tailed tits in the garden, and on Wednesday a sparrowhawk flew over and something that looked like a garden warbler appeared, but I'm not sure about that.' Now more likely are quick-fire texts and Instagram posts: an instant sharing of experiences, photographs (but making few lasting memories?). A more knowledgeable friend is always a useful bonus, but who has one of those? There may not always be one about, and you can't just conjure one up. Try a county bird club – the focus for local good and experienced birdwatchers – and the nearest RSPB group, which will be more about helping people meet each other, learn a bit more, and naturally encouraging them to support the RSPB (which deserves and needs your support in equal measure). My first proper RSPB job was to help these groups, which my boss, Trevor Gunton, an RSPB hero if ever there was one, was responsible for creating. Or just chat to birdwatchers you meet outside: most are a sociable lot, although birdwatching itself can, admittedly, be a very isolated, solitary activity.

For most of us, we watch birds simply because we like them. We meet beings of wondrous variety and endless activity – and they can fly! And it is all basically free. A bit of learning will enhance their gift of company and entertainment with understanding. They may add a bit of colour and life to our surroundings, or they sound nice (if not always, perhaps, gulls on the roof at 4.00 a.m.). Individually, or collectively, they can look beautiful, impressive, dramatic, or delicate, exquisite, fragile and vulnerable. Some *might* look funny: I tend to resist laughing at them because I think they deserve better than that. I don't think of a puffin as 'amusing' despite the media image. And while any given species has its own character and characteristics, all birds are individual beings.

Soon birds will be linked with time and place: great memories of moments or days or holidays, or wonderful places we've visited where the birdlife played a major part. Knowing that a swift has come from Africa and hasn't set foot on the ground since it was here last year adds a bit of wonder, too. Birds and places will be linked with family and friends, holidays and days out: great memories. Yet it is strange what memory does and doesn't do. I can recall ordinary, simple things from decades ago with enormous pleasure: yet, when looking back through my notebooks, I can now find long descriptions of rarities that I can't recall even seeing. I'd forgotten a white-tailed eagle in Suffolk; I remembered a sociable plover but forgot about a couple of others; likewise with blue-winged teal. Some things leave an indelible impression, others a transient little dent, but I find that writing things down generally helps reinforce the later recollection.

My background education in geography also, I believe, helps me understand a bit more about the natural world. Biology meant cutting up rats and frogs, so I dropped that very early, before O-levels. But I still have a bit of a biological slant when looking at things, which you need to have, especially if conservation becomes a concern. That adds to the fascination of birds. How does a tiny warbler fly to Africa and then return a year later to the very same tree? Anyway, why does it bother? Think of everything that makes a human keep going: brain, nervous system, liver, lungs, kidneys, stomach, intestines, immune systems, heart and blood and arteries and veins, inner-ear complexities to keep you balanced and sort out the sounds, eyes and optic nerves to see the world and understand it. The tiniest mouse has the same... the smallest warbler is so much better at many things than us, its reactions so much quicker. Even a butterfly can migrate across Europe. How does it know? Why does it decide to get up and go? Those bats in the evening, using their sophisticated echolocation systems, something entirely beyond our own physical capabilities. Watch a tiny insect zip along a hedge then climb up and away across the garden, in a matter of seconds: in proportion to a six-foot human, what fantastic speeds it must achieve! Such forces of acceleration and deceleration, probably while going sideways. And how ever do those moles under my lawn find their way about? The natural world, from the smallest beetle to a distant black hole, is exciting yet difficult to understand or comprehend.

The way we react to the world around us is fascinating too. Is it too strong to say that many people look at the world through the television screen? This thought goes back a long way. When I was a student in Swansea, a huge wooden cabinet appeared one day in the coffee bar. A colour television! No-one had ever seen one of those – a small crowd gathered round. I clearly remember one comment, 'Look at that red! That bright red, what a colour.' And equally clearly, I recall my own mental response: yes, bright red, but it's only a garage door... there have been red garage doors out there for decades and you've never looked at one before. Of course you must appreciate the wonders of television, and natural history films in particular have reached an incredibly high level, but do people perhaps look at a wonderful sunset on the box but never notice the real thing outside?

Sadly, many people who get out into a 'green space' for the benefit of their mental health just stick in the earphones and run, or cycle, and don't really have any kind of reaction to the world around them, or at least not in any detailed sense. But there I go again, judging others... I must stop. It is up to them. We may be stimulated by beautiful things or simply ignore them as ordinary, everyday, background patterns. Or, realistically, both, depending on our mood at the moment. But whereas I will stand and look at the moon, or stop to admire a rainbow, I know most people simply don't take the least bit of notice. Children look at rainbows; why not everyone else? I can't *not* stare at and feel moved by the brilliance of Venus or Jupiter whenever I see them, but most people are unaware.

Life is immensely complex, and the world as a whole even more so (let alone the universe). There are millions of grass leaves on a large lawn, which have millions of dewdrops reflecting the early-morning sun, each with sparkles and intense rainbows of colour, tiny spectrums (spectra?) of brilliant hues. Within a lawn there may be so very many eggs, cocoons, larvae of this and that, tiny insects, slugs and snails and spiders, all of which have their own complicated and brilliant nervous systems, digestive systems, remarkable eyes (complete with whatever is needed to make them work and process the information they provide), extraordinary jaws, legs, wings and other means of locomotion, senses and reactions to external events. Or maybe they're just insignificant ants or gnats, to be trodden on, swatted without a thought, smacked against a car windscreen or snapped up by a robin. A spider to be squashed. And the lawn can just be ignored most of the time; indeed, it *must* be ignored most of the time, because you *can't* simply worship it all day long... it's a bit of green.

Or we may drive by a bit of woodland, a fantastic ecosystem built up on such minutiae with seemingly infinite numbers and variety of flowers, butterflies, beetles, birds, leaves that glow green with chlorophyll or turn red and gold with accumulating sugars as autumn approaches. Or, as a rule, it's just that clump of trees across the field.

Even the most mundane, everyday bit of life is amazing. The temptation for a writer from time to time is to try to describe all of this, to react to it in some way, to try to encourage the same response in the reader, to impose some reaction upon them. In fact, everyone will respond in their own way. Producing a bird book can suck you (me) in to enthusing about the way a big bird of prey tilts and turns in the air. How its upcurved wingtip feathers, the outermost curling more than the next and so on in sequence, change the overall appearance as it turns, from a broad, fingered, rounded wing to a pointed one. Changing from a long, oblong slab to something more elegant, with bulging secondaries, an S-shaped trailing edge, shapely wingtips caressing the air. How the tail reacts to air currents, the alula (thumb) moves in and out to give extra control over airflow and turbulence. Trying to enthuse, trying to show off or boast about what you know (which only ever reveals how much you *don't*). I'm just trying to *appreciate* things properly, because surely they deserve it.

Blackbirds on the lawn beneath the apple tree, with strong sunshine casting dark shadows. Such complexities of light and shade are always fascinating, however commonplace the setting or the bird.

Yet I can still at least say that something is beautiful and remarkable. Perhaps a wave breaking on a beach, glaucous and transparent in that last, curling, over-toppling moment, the sun reflecting off myriad facets on the sculpted surface of the underside of that glistening curve. I might clumsily try to paint it, more likely photograph it. But it is just a wave. Splash. Wait for the next.

The lawn with its multitude of reflecting diamonds is a just a lawn, the bit of grass where you throw out crusts for the birds. The blackbird is just a bird, hopping along with its bright eyes alert and expectant, reflecting the sky and the trees, the garden shed and the roof if you are close enough; its wings drooped, tail raised as it pauses and slowly lowered again as it stands still, its legs flexed, the tension in its neck giving just that right kind of poise that *defines* a blackbird.

Just a bird, but fantastic: black, but beautifully black, lamp-black, carbon black and inky-black and soft charcoal-black, and brownish black and bluish black – even pale black, if you can have such a thing, when the morning light catches it and filters through the minutely raised feathers. And somewhere within it, the capacity for producing one of the most musical of all bird songs. And those lovely softly mottled eggs. Indeed, the more I think about it, the more it becomes a favourite bird, giving more pleasure, more of the time, than most.

Just a bird. But I like it. I hope you do, too.

WHAT TO DO WITH YOUR INTEREST IN BIRDS

Birdwatching has no rules (except that *the welfare of the bird comes first*: look at the Birdwatcher's Code at the back of the book). Do what you want, so long as you don't trespass or do damage. But who knows where you might end up?

One aim of this book is to recount real birdwatching experiences of all kinds: a walk in the countryside, an hour in the garden, a long-distance journey to see something new – all the different ways in which I have enjoyed birds over the years. You don't switch off. Birds are there outside the window, on the way to school or to work, over the car park by the shops, a reflection or a shadow. Even while I'm writing this there is a sparrowhawk doing a recce outside... once you start, you never stop.

I feel sure my education in art and geography has given me a particular view on birds, their habitats, the landscapes they inhabit, the way they look in changing seasons and ever-changing light. You may be knowledgeable about or interested in climate and the effects of weather on bird behaviour; you might check the excellent weather-forecast maps after the BBC news to predict what tomorrow might bring, bird-wise. You may almost subconsciously think about the geology of places you visit, and see what birds might be there in the particular vegetation types growing on soft underlying chalk or limestone, or dark, unforgiving millstone grit or slate. We often had family trips to the Peak District, so clearly divided into the White Peak (limestone and green grass) and the Dark Peak (dark grits and heather moor). A bit of geography – *real* geography, I mean, physical geography, not place names or flags – informs your understanding of habitats, which helps you predict the birds and *understand* them a bit more. You might study biology or ecology; or birds might stimulate you to look into these subjects. All of these things just tend to build up slowly over the years and become routine, subliminal.

You could record bird sounds and analyse sonograms (can you count a bird if you have only heard it?), or get a camera and long lens and take thousands of photographs (can you count a bird if you see it only on a photo later?). It is no longer a case of saving up for a 36-exposure 35 mm film and taking pictures carefully at 50 pence a throw. A photographer in a local hide is not unkindly known as 'Machine Gun' for his habit of keeping his finger firmly to the button. Or just use your phone: the pictures (and videos) they take can be incredible, and you can get all kinds of adapters for telescopes and so on. Maybe you can draw and paint. That would open up another couple of books. Or perhaps write: Dylan Thomas included plenty about birds in his stories and poems; Thomas Hardy wrote one of the best evocations of a songbird in his poem *The Darkling Thrush*. My favourite, Dorset dialect poet William Barnes, was always writing about rooks 'fleeing to their elems'. The elems (elms) have long gone. Nature writing is currently very much in vogue. And Vaughan Williams produced the perfect rendition of a singing skylark in everyone's favourite, *The Lark Ascending* – maybe you will write more bird-inspired poetry, prose or music.

You may well wish, though, to add something to just looking at birds for enjoyment with no aim in view: add a bit more purpose. That is easy enough

Some people see Thomas Hardy's Darkling Thrush *as miserable and pessimistic, but its ending is surely a joyous sign of hope and optimism for better times ahead: 'there trembled through / His happy good-night air / Some blessed Hope, whereof he knew / And I was unaware.'*

to begin with, at a simple level (the kind I have managed myself without going very much further, most of the time). One easy thing to do when you see birds, especially at a place you visit regularly, is just to *count them*. Numbers add interest to any list, give you a better idea of what you saw when looking back, allow you to send in your records to a local bird club or online records system, and help to reveal what is going on from day to day, season to season and year to year. Not just increases and declines, but seasonal movements become obvious.

Counting is the simplest, most basic element of survey work. Bird surveys can get far more complex. Counts are invaluable, but you need to add an element of discipline. It is of little use recording, say, a single high count of meadow pipits, without some context at least, such as a regular series of figures or at least an informed summary of variations, so that the one high figure takes on some significance. Counts need to be frequent and will require some sort of basic analysis to be of much use, but that is all you need, to begin with. Add that bit of purpose, that desire to illuminate the changes, the movements that are always going on.

And do, please, submit your records: just look online for the local bird club or bird recorder or your county 'goingbirding' website. Years ago, we used to go through our notes each new year, to write out a summary of records to send in to the county bird recorder. These would be on little record slips supplied by the bird club. Later, when I was one of the co-editors of the bird report for a short while (for the admirable West Midland Bird Club), we used to divi up record slips according to species or area and have shoeboxes full of them, all arranged in species/date order. From these we would try to produce a useful and meaningful summary. These old published reports still have interest and value. I always thought the vast amount of material contained in county bird reports was grossly underused. A different function, now, is to bring back memories of those days and the places and people involved: mostly known by recognisable sets of initials of the observers, put in brackets after any significant record.

There are also nationwide surveys organised by ornithological and conservation bodies that you can become involved with, given sufficient experience. These monitor common birds in town and country, and along waterways; numbers of nesting pairs in the UK's heronries; the numbers and distribution of wintering waders and wildfowl; there are also periodic counts of wintering gulls. There is plenty to join in with and plenty more that you can start up for yourself. You do not, of course, *have to* become a dedicated survey worker; you very likely won't have the time. But it is a possibility, should you wish to consider it.

There are, of course, volunteering opportunities, and some professional possibilities with organisations such as the RSPB, BTO, Wildlife Trusts and others.

These will be limited and relatively infrequent – but why not make yourself known, if you are keen? Volunteering might be a way to start on the precarious career ladder. Somehow, I bumbled my way into a career because other people noticed me. I should have been more active in trying to get myself known, but I did once speak to Trevor Gunton at an RSPB event long before I got to know him, and once made a phone call to Roger Lovegrove in the RSPB Wales Office, asking if there was any chance of a job. A few years later, he phoned me. Still fancy working for us? Yes please!

Joining the RSPB is a good way to help birds; joining the BTO (British Trust for Ornithology) is a good way to learn more about them and to find practical and useful ways of getting involved. Look at their websites. For an international perspective, look at BirdLife International, and All About Birds, the wonderful website that links the recent *Handbook of the Birds of the World* with the Cornell Lab of Ornithology. It is through the BTO, too, that you can be trained as a bird ringer. This is a specialist pursuit, involving catching a bird (or handling a nestling) and putting an individually numbered ring on the leg. If it is later caught by another ringer or found dead, the details will give an exact record of two points in its life history when it was at a precise location at a particular date. It might be just down the road next day, or in a remote foreign country years later. Ringing (and the subsequent development of satellite tracking, even for small birds) has revealed an immense amount of information that could never have been accumulated in any other way, regarding longevity, population dynamics and migration.

BUY A GOOD BOOK

Okay, I know you can download the app... But books have a place. Many of us older birdwatchers have an attachment to our early books, too. There may be rows of them on the shelf, even if they are never used. Mine are battered and bruised, some covered in brown paper, bindings falling apart. I'm not going to give detailed advice to help you buy a good book now, although there are so many excellent ones around, but the history of bird guides is an interesting subject in itself and, one way or another, I have somehow crossed paths with a few of the field-guide writers in a very small way.

There were early guides to seabirds, and books such as W. H. Hudson's *British Birds*, a bit of a cheek from a man recently arrived from the Argentine pampas. If you look it up you will find several different dates cited, but 1895 was the first edition, with a number of improved and expanded versions published into the 1920s, which were more useful guides and especially good at evoking bird songs. The first comprehensive bird 'field guide' dealing with *identification* and nothing much else, though, was developed in America by Roger Tory Peterson. He then produced a European guide using a similar style and approach, together with Guy Mountfort and P. A. D. Hollom: the 1954 Collins 'Peterson, Mountfort and Hollom' guide, which quickly became a best-selling classic. Following these came guides by Richard Fitter and Richard Richardson, and then Heinzel, Fitter and Parslow, before a multiplicity of titles, good, bad and indifferent.

Around 100 years ago there were good, serious bird books, but nothing you might call a field guide, or pocket guide, and certainly nothing cheap and easy to get. Horace Alexander, a great British birdwatcher, watching birds from the very end of the 19th century, said that as a child he was given the seven volumes of Dresser's *Birds of Europe* – but a grown-up had to make sure his hands were clean before he was allowed to use them. His brother had Howard Saunders's *Manual of British Birds*, another multi-volume set, the standard until the 1930s, before getting W. H. Hudson's single-volume *British Birds*.

T. A. Coward was a pioneer who produced excellent essays, with just adequate identification material, in *Birds of the British Isles and their Eggs*, in 1920. It was rejigged in the Wayside and Woodland series (published by Frederick Warne) that I would much later borrow from the library.

A huge step came with the great Witherby *Handbook of British Birds* of the late 1930s, a wonderful set of five substantial books – these had breakthrough texts on identification and behaviour by B. W. Tucker. He sent the draft texts, written out by hand, to that same Horace Alexander for his comments. In a very tiny tenuous link, just to allow a bit of name-dropping, although of course I could never have known anyone involved in these books so long ago, I did once get a letter out of the blue from Horace Alexander himself, about inland migration of kittiwakes – a letter to me, from someone who helped write the Witherby *Handbook*!

Strangely enough, in 1936, before the *Handbook*, Witherby had published a different guide, written by someone who I didn't really know but did at least meet, through work, Max Nicholson. Max, a particularly deep thinker in conservation, was ahead of the game in so many ways. He lived in Cheyne Walk in Chelsea, where the footballers and rock stars live. He wrote the text for *Songs*

My old books still stand, battered and well-thumbed, on my shelves, alongside newer volumes. Some still mean more to me than any of the current guides, even though they may be far less useful. Now and then I pick one up just to remember those early days, working things out as I went along.

of Wild Birds, a set of 78 rpm records recorded by the amazing Ludwig Koch, with illustrations – the first multimedia guide. Koch was German, Jewish, but by a stroke of good fortune escaped to England and became famous for his remarkable wildlife recordings.

But the Witherby *Handbook* of 1938 was a set of five books: you couldn't take them out with you. You had to write notes and make sketches while you watched the bird and come back home to compare them with the book, or rely on your memory. This was the way that birdwatching books recommended anyway, for a very long time – but you could easily miss the very thing you needed to take note of. And in that case you just had to go back and hope the bird was still where you left it the day before.

The first bird books that told me how to tell one bird from another would have been my brother's Ladybird books of British birds – not the later ones by John Leigh Pemberton, but the originals from 1953 and 1955, written by Brian Vesey Fitzgerald with paintings by Allen Seaby, with a third one illustrated by Roland Green in 1956. It's remarkable how books, just like music, stay with you from such an early age. I remember pictures of bullfinches and yellowhammers from these Ladybirds, and these two are still among my all-time favourite birds, and with two of the best names. Roland Green also wrote and illustrated *Wingtips* (one of our school library books), a guide to birds in flight, in 1947, and he had much earlier illustrated simple bird books by Enid Blyton in the 1920s, as well as the much more substantial *British Bird Book* by W. P. Pycraft and Theodore Wood in 1921. Who knew that Enid Blyton of children's story fame was a pioneer of the bird-book genre too?

A book I learned a great deal from in the 1970s had the relevant bits of the Witherby *Handbook* distilled by P. A. D. Hollom into the *Popular Handbook of British Birds*. My well-used revised fourth edition of 1968 had line drawings by Donald Watson, Richard Richardson, Ian Wallace and Peter Hayman, all great in their own ways, all of whom I met or, in Peter's case, knew quite well. Watson produced scraper-boards of great richness and intensity, Hayman more measured and angular drawings, while Wallace managed to get a multiplicity of lines and scribbles to form a smooth, uniform surface; Richardson just did a minimum of lines and got the thing to perfection! He is impossible to imitate.

Incidentally, the word 'handbook' became synonymous with multi-volume sets with masses of information, not small books to be held in one hand – or put in the pocket. Identification books became 'field guides': the name itself probably put a lot of people off, thinking 'that's not for me'. Over the years I've been involved in tracker guides, glovebox guides, field guides, pocket guides, easy guides – all kinds of things.

What can be better than a splendid bullfinch? What better name than redwing? These things appealed to me when I was 11, and still do. I used to share an office with Carl Nicholson, who said I drew birds standing in grass because I couldn't do their feet. This is for you, Carl.

When I was small I used to go on the bus to Lichfield with my Mum, and there was a nice little bookshop where I could see, smell and feel real books, and I would trot across to look at a shelf full of little white ones with coloured bands and numbers on their spines. They covered anything from astronomy and weather to music and cacti, from ships to freshwater fishes and mammals – and of course birds and birds' eggs. The Observer's books!

The Observer's Book of Birds used the same pictures as Coward – great paintings, but not always of a field-guide sort of style and showing just one or two plumages. But it was certainly a book for the pocket – the *Observer's Book* may sometimes have been in mine, though I suspect it was usually left at home. But I was using a helpful book at a young age. And I was making notes – the early ones have gone, but my first surviving bird record is of a jay in Sutton Park, when I was 11. It was probably the *Observer's Book* that helped me tell what it was – I can still see it now, flying across a lake, while Dad rowed us across in a boat. Funny that the first bird book I ever wrote was a complete rewrite of this one.

But I needed something better and more comprehensive to get beyond basic beginner stage. I was able to twist what I saw into what I wanted to see with the

Observer's Book – I remember some bright finches on a hedge, tawny coloured with big beaks and white wing patches – I *wanted* them to be hawfinches and wrote them down as such, but really I think I *knew* they were chaffinches in the winter sunshine. I needed to be a bit more disciplined in the way I looked at things. And I needed something that would open my eyes to other birds I knew nothing about, to make me aware of the possibilities – so I could do some homework.

I chose the Collins *Pocket Guide* written by Richard Fitter and illustrated by Richard Richardson, originally published in 1952. I could have got the Peterson *Field Guide* of 1954 (with many subsequent impressions), but I think I chose the other one because I thought the illustrations were more realistic – more natural, I suppose. Richard Richardson was a fine artist, unmatched in his line drawings, and the pictures in the book were not all in the rigid, slightly flat style we tend to associate with guides. They looked more natural, nicely composed, nicely compared but not rigidly sticking to the same repeated pose. The text was good but oddly structured, using habitat as the main division – land, waterside and water, which never really works – then size, large, medium or small. But it was a good book and got me through several years.

I thought the Peterson illustrations in the other guide were a bit flat, unreal, even a tiny bit cartoon-like – I remember the strange forward-leaning moorhen and coot. But, looking back, I see how real they are – people talk of Peterson's pictures

My first surviving bird record is of a jay flying across the lake in Sutton Park.

as patternistic, and his early ones may have been, but the European guide is a fabulous piece of work. Many of the plates are just *beautifully* done.

Roger Tory Peterson created his first guide to the birds of the eastern USA in 1934, then a guide to western birds, before taking on Europe after the Second World War. He was credited with 'inventing' the modern field guide, but apparently Edward Wilson had been working on something before he fatally went to the South Pole with Captain Scott. This work was expanded and completed by W. B. Alexander, who produced a real guide to the birds of the ocean in 1928. I bought a copy of the 1963 edition in 1975, because I thought I should have it, but I never used it, and it was not a very helpful guide by modern standards – though still a massive breakthrough at the time, when there was really nothing else. And of course, W. B. was the older brother of H. G. (Horace), the very one who wrote me that letter!

My boss at the RSPB, Trevor Gunton, with my wife Marcella, used to devise and run all kinds of RSPB events. With the help of Crispin Fisher at Collins publishers, Trevor found out that the third edition of the Peterson *Field Guide* was being reprinted in 1979 and invited Roger Tory Peterson, together with Guy Mountfort and Phil Hollom, to celebrate it at the RSPB Members' Weekend – there would be a signing session and Roger would give a talk. Crispin's father, by the way, was James Fisher, who not only wrote books with Roger Tory Peterson, but also produced a set of small identification guides for British birds, *Bird Recognition*, illustrated with black-and-white drawings.

Anyway, I knew nothing of these plans until Trevor, out of the blue, told me that he'd got Peterson to do a short lecture tour, and I was to drive him and his wife Ginny around the country, deliver him to the hotels, and introduce him at his lectures. It was, of course, an enormous privilege. Roger's talk included the story of the first field guide – which gave rise to the whole series of Peterson guides to practically everything. His illustrations showed birds side by side in similar postures; emphasised the pattern – flight patterns, for example, were in black and white – and picked out the important marks with neat little pointers, a unique feature at the time that came to be rather grandly known as 'the Peterson system'. The text had important features highlighted in italics, which we all copy now. The same style was used in the European guide, first published in 1954, and the original text was thoroughly updated by Ian Wallace and James Ferguson-Lees for the enlarged edition of 1966. Revisions of Peterson, Mountfort and Hollom brought it up to date and it became indispensable for many years.

It was not, to be honest, a big deal, but the illustrations were scattered through the book, neither next to the relevant texts nor bunched together. It took only a few seconds to find what you wanted, but still, you had to flick through the pages to match picture and text, and other people saw a way to challenge the Peterson guides and put the text opposite the illustration for every bird. In America the *Golden Guide* by Chandler S. Robbins put text and pictures opposite each other on a spread. Richard Fitter saw it, and knew something similar would go down well in England.

When I was editing *Birds* magazine I could for a while, up to a point, decide what went in, as you might expect an editor to do, without too much worry about RSPB brand or image or mission statement or whatever. I had an ambition for years to talk to Richard Fitter about the development of the identification guide in Britain, in parallel with the Peterson guides in the USA – Richard was really as much a pioneer here as Roger was over there. So I thought I would just ring him up and see what he thought. He was delighted. When I went over to his house one day in 2005, not so long before he died as it turned out, he said his carer wasn't there at the moment but if I'd just sit down, he'd make a cup of coffee and tell me all about it... which he did.

He had piles of *Birding World*, *Birdwatch*, *Bird Watching* and *British Birds* magazines on his desk, bound sets of *The Ibis* and *Bird Study* and various botanical journals, a neat shelf with a complete set of Ordnance Survey maps, all the books you could think of, letters and papers scattered around his typewriter, and he was busy revising one of his wild-flower books and starting on another – he was 92!

He told me about his years of work with the Fauna and Flora Preservation Society and his politicking on behalf of wildlife conservation – and began to arrange a day out for us to go and look at some orchids. My esteem for this man went sky high.

He had met Richard Richardson when he was a young boy – now, as I have since realised from my own experience, this is a good example of field-guide text pitfalls – when who was a young boy? Richard. They are both Richards. When Richard Richardson was a young boy. Ah, right. Richard (Fitter) used to call Richard (Richardson) Dick to avoid any mix-up. Later he (Richard) asked him (Dick) to illustrate his bird guide, and then a guide to birds' nests. And Richard Fitter produced Collins field guides to wild flowers, as well. But on a visit to America in the 1960s he saw the *Golden Guide* and said to Collins, if we don't do something like this quickly, someone else will – and so came the next Collins guide, *The Birds of Britain and Europe* of 1972, soon universally known as 'Heinzel, Fitter and Parslow'.

This new guide finally had all the birds illustrated on the page opposite the text, as well as the maps. But the text was sometimes a bit too brief, and the coverage initially seemed a bit too broad. I remember being unimpressed at first when I eventually bought it – I had no money to buy books at the time so I had to be choosy. Who wanted to know about the rare African striped crake or some odd wheatear in Israel? A few years later, *I* did, and this was still the book to do the job on any Western Palearctic holiday for a very long time afterwards. It was not so widely publicised, but years later this book was revised, with many European races thoroughly covered. The revision must have involved a great deal of work, and made it a very good book.

To be fair, Richard and Collins had in fact been beaten to it, by the *Hamlyn Guide* of 1970, written by Bertel Bruun and Bruce Campbell and illustrated by the American artist Arthur Singer. Heinzel, Fitter and Parslow came out two years later, in 1972, but it was generally believed to be much the better of the two.

So the text faced the pictures. This is fine, but it does tend to squeeze the space for words. It means less information – but it puts the writer on his mettle and forces him to write tighter, more efficient prose. This can be tricky, as I know only too well (and as the two Richards story above also shows). You can, for example, say 'X is like Y; larger bill and browner head distinctive'. Does that mean X has the larger bill and browner head, or Y? There are endless pitfalls around ambiguous text. It's a bit like the sign that says 'Police Slow' – so you think great, I can drive a bit faster and they'll never catch me. Or you can have something like I found in the Pyrenees, when I sat with an open book beside a gentian: 'There are two similar species, best told by this one being slightly larger than the other.' Erm... I can only see the one! A seemingly good guide can suddenly become a bit useless.

Also, there was a trend to replace long words with symbols, or icons, or abbreviations, putting a lot of information into a much smaller space but often making it more or less incomprehensible, and off-putting for beginners.

The new Hamlyn and Collins guides spawned a lot of competitors. Suddenly, the birds in your pocket were big business. I remember John Gooders phoned me to complain about a review of his Kingfisher guide in 1990 – 'Come on, Rob, I have to make my living with these things.' Fair point. I'd said, I think, that the text was a bit rigid, and that 'head white, neck white, breast white, back white, wings white, tail white' was maybe not necessary when describing a swan. Actually, it was a very good book (as you would expect from Gooders), illustrated brilliantly by Alan Harris – not only authors but talented new artists were producing good stuff. Alan Harris has produced some wonderful field guide material.

Only recently did I learn that Peter Hayman was in the army, and was shot in the leg, so had to spend months lying on his back recuperating, somewhere in the Middle East or Mediterranean. So he studied the birds flying overhead, and became a bird of prey expert, and used to give drawings to people travelling abroad so they could tweak the postures and patterns and help him to get things as right as possible – he did the same for me for African vultures. Peter is a great artist anyway, and trained as an architect, and his precision and dedication brought about his own style of illustration. He produced a really innovative book for Mitchell Beazley in 1976, *The Birdlife of Britain*. He based his illustrations on complex series of measurements of museum skins combined with his own knowledge and study of other references.

In 1979 this spawned the Mitchell Beazley *Birdwatcher's Pocket Guide*, truly a tiny guide that would genuinely slip into any pocket yet still had a comprehensive coverage of most European species. Later, Peter specialised in waders and illustrated the breakthrough Christopher Helm guide to shorebirds; he would have done seabirds, as well, if Peter Harrison had not beaten him to it.

A friend of mine was involved in producing the first Mitchell Beazley book, and tells stories of having to lock Peter away in a room for a couple of weeks until it was finished. In the same vein, Richard Fitter said that Richard Richardson found painting more than 100 detailed plates exhausting and used to try to get a break by walking to the shop for a packet of cigarettes. Richard got wise to this and would send his wife to get them instead, so Richardson would get on more quickly with his work – which, by the way, was apparently largely done from memory! I admire any artist, however good or bad, who takes on a job like this.

This raises another problem with these guides. An artist might take years to do such demanding work. Who will pay him? Can the book's economics stand the bill? Peter went home to tell his wife he'd signed a good deal to do the shorebirds of the world, and I imagine Dorothy thought fine, that fee might pay Sainsbury's for a few months. Peter said, well, actually it's for two years, full time...

Peter also created a series for the RSPB magazine called *What's That Bird?*, subsequently made into a large-format book filled with illustrations. Mike Everett wrote the text. It was really just a promotional thing for the RSPB, and I used to give it away to anyone who joined the RSPB at film shows, but it had some fine work by Peter, especially the beautifully accurate finches, buntings and birds of prey.

Determined to produce something of lasting worth, Peter Hayman negotiated a new deal to do a revised Mitchell Beazley guide. He asked me to write the text to help split the workload, but all the new information came from him, which is why I can happily say how good it was without being immodest – I really mean how good Peter's work was. There was a good deal of new and innovative stuff in *The Complete Guide to the Birdlife of Britain & Europe* of 2001, which I regard as a modern classic.

Peter measured everything, mostly from specimens preserved in spirit at the museum – he had to dry them out with a hairdryer first, then they would become supple, just like a fresh bird rather than the stiff dried skins in the drawers – Peter said he was often drunk on the fumes before he started. He also had birds x-rayed so he could study skeletal structures, sorting out quite a few relationships that way, before people started using DNA – his unpublished seabird research was really ground-breaking. This book again led to a new Mitchell Beazley pocket guide – I still see it around, and it is still a good one, if a little bit too small and squeezed. But all in all, I think Peter, lauded as he was, is still underrated and his major works bear comparison with anything you care to mention: I still refer to his illustrations time after time.

I'd been involved in a few things before I worked on that book with Peter. I'd rewritten the famous Observer's Book for Frederick Warne – I was paid £180, I think, and had royalties! We used to go out for a meal every six months to spend them, but I would have done that book for nothing. Then I did a book for Usborne, illustrated by Trevor Boyer, another very fine artist – this sort of work gets you to meet all sorts of interesting people. I began by supplying typed sheets, moved on to floppy disks, then Syquest disks – very fragile things, like spinning plates balanced on a stick, if anyone remembers them – then CDs and DVDs. Now everything is done by email and assorted attachments.

One more great figure in the field guide story is a rare man, in that he both wrote and illustrated the books to an exceptional standard. At the tail end of the 1970s, Lars Jonsson set off around Europe from Sweden, and produced a set of small books that covered Europe's birds. They were beautifully done, and his birds fizzed off the page. Sometimes, I thought, they were a fraction more 'Lars Jonsson' than the bird itself, but the guides rightly received rave reviews.

But if you are trying to convey the bird perfectly, how come it is possible to say 'that's an Alan Harris painting' or 'that's Peterson' or 'that one's by x, y or z'? Surely they should all look the same? If there is too much individual style, isn't that a bad thing? Not necessarily, perhaps: Lars Jonsson's work is instantly recognisable and consistently brilliant. When he put everything together in one volume, revised many plates and added new ones, the resultant *Birds of Europe* (1992) was sensational, and still is.

Lars pointed out that Peterson's first American guides took people from shooting birds and comparing them with handbooks to capturing the sort of image they could see with binoculars – but modern field guides now need to go back almost to the bird in the hand, as people now see them ultra close-up with telescopes or on their smartphone photographs. He said that in the 1970s he could get away with saying just that immature plumages of skuas were difficult to tell apart, but by the late 1980s they all had to be described in great detail. Since he said that, it has become even more like examining birds in the hand – specimens – all over again, with the great increase in digital photographs, freely available on the internet. Count the emarginations...

Lars takes things to another level in his identification papers on smaller groups of birds such as skuas, divers and gulls. He can capture the essence of them all on paper, in a way that goes far beyond the field guide – he gets into the character of the bird in a different way, almost the spirit or the 'being' of the bird, through incredibly accurate painting and prolonged close study of living birds. Perhaps it becomes a bit personal – if we can't see what he is getting at, maybe it can't be described in words – if you don't get it or can't see the nuances that he can, it doesn't work. As we aren't anywhere near as good as he is, maybe we just can't always easily understand what he means – but his work is just extraordinary. Here are some gulls, and here's a semipalmated plover compared with a ringed plover – subtle, but it's all there. I almost got Lars to do a series in the RSPB magazine, but he wanted to start with the robin, and couldn't do that until he came over to England again, because English robins looked different from Swedish ones and he wanted to get them right.

And of course that brings us to the current leader of the field, the superlative 'black' *Collins Bird Guide* by Lars Svensson, illustrated by Dan Zetterström and Killian Mullarney, first published in 1999 and now in its third edition. When Killian first came on the scene, practically unknown in England, I remember Ian Wallace said his field notebook was the best he had ever seen, bar none. The Collins guide took things a further step forward, and the illustrations were so good that a large-format version – nothing like the bird guide in your pocket now – sold well just because people wanted to see the pictures more clearly. It was both a field guide and a coffee-table book. And later it became an e-guide.

It is, though, at a different level from earlier books. I well remember, when I was still at school, seeing a small brown bird – my cousin said it looked a bit like a chaffinch with just one wing-bar instead of two. Richard Fitter's first pocket guide led us to pied flycatcher – a bit like a hen chaffinch without the white shoulder patch, it said. Yes, spot on, and just exactly right for us. The Collins guide, which is of course brilliant, leaps in at a higher level altogether, almost assuming you know it's a pied flycatcher already but telling you how to separate it from a collared or semi-collared. How do you get a book to work at both levels, if such a thing is possible?

And anyway, could there be another way to do this? How about using photographs? That was a dream for many years – but the photos simply weren't there. Alan Richards, ever the optimist and pioneer with his Aquila Photo agency, published a magazine on birds, *The Midland Birdwatcher*, long before any others, and created a nice book on *British Birds* for David & Charles in 1979, with excellent photos and some of my little line drawings, but only one plumage per species was shown. Roger Philips produced a large-format photographic guide to butterflies and moths for Pan Books, and in 1980 Nick Hammond and Mike Everett followed it up with a companion bird guide. Most species had just one photo, and I added a black silhouette for each.

Paul Sterry was an early entry into the photo-guide stakes and still produces good things, including the recent BTO guides. Years before, I remember two big Italian books of photos of European birds, neither small enough for the pocket nor comprehensive enough for the shelf – more ornamental than useful. The photo guide was a difficult enterprise, except perhaps for small groups of species. How could you get the necessary coverage – male, female, juvenile, winter, summer, upperside, underside, perched, flight... an artist could do all this but, still, photographs could not match it, even just a few years ago.

I was lucky enough to be asked to write the *RSPB Birds of Britain and Europe* guide for Dorling Kindersley, published in 2002. Despite limitations of available photographs, DK included several new ideas, and still revise it to keep it up to date.

Richard Crossley then came along with his *Crossley ID Guides* to North American birds, which suddenly changed things – now there were several, sometimes tens of pictures on each page, showing birds in all kinds of plumages, near and far, doing all sorts of things. A bit of reality, but sometimes in an unreal design – some pages seemed to have birds scattered everywhere, others looked to me a bit like a set of birds in a glass case. But his books are a real tour de force.

Photos can, sometimes, look a bit like stuffed birds. And they can be misleading. A field-guide artist can exclude extraneous lighting effects, but a photo might not – you might get green reflections from a leaf, blue reflections from water, blue shadows and orange highlights in strong sunlight on snow – the sort of thing you learn about at school with the colour wheel, the kind of effects the Impressionist painters tried to capture. An artist, or a good photographer, can use such effects to produce beautiful images – but a field guide doesn't want that, it needs effectively flat, evenly lit, natural-colour pictures of the bird.

People would send paintings to the RSPB magazine, often obvious copies of photographs – so, you might get a painting of a whitethroat with a bright blue chin, or a pied flycatcher with a silvery back. These were effects of light and shade captured by the camera but replicated completely out of context, with no understanding of the situation, to give entirely inaccurate results. There's no point making a field guide with photographs that give the wrong impression of colour to people who don't know the bird and can't interpret them. Even shape can be tricky – in the DK guide I remember I insisted on having an out-of-place tail feather photoshopped out of a Cetti's warbler picture because it gave the impression of a forked tail. That same picture is still being used, with and without the unruly tail feather.

Yet when writing the recent WildGuides/Princeton University Press guides to birds of Britain and birds of Europe, one of the great joys was having a password for the brilliant Agami photo agency images, as well as for the personal files of David Tipling, Hugh Harrop and Andy Swash, and scanning through thousands and thousands of stunningly good pictures. As with Peter Hayman's paintings, I can rave about how good the photos are in the books, as I didn't take them.

But how do you make a field guide that works, especially for beginners? You know for a start that half the people will not read half the words. When I became involved in writing these new photo guides, I wrote twice as many words as were finally used – so perhaps that means you will read all of what's left, but I doubt it. I wanted to say a lot more about how birds moved, how their shapes change from different angles and as they move in different ways – all the subtleties that you learn over the years. Like the goshawk, for example, which I have come to know better in recent years – all the subtleties of shape, wing set and so on as the bird does this and does that, soars, flaps and stoops into a wood.

But in the end, you can't really do all of that – the beginner has to learn these things for him- or herself, and it is the best way to do it. Identify the bird, then get familiar with it, get it into your head.

There were many things left out of these guides that could have been useful. Swallows, house martins and sand martins, of course, feed in the air, catching insects. Obviously. But I have seen flocks of them feeding in foliage on tall treetops and more than once large flocks – hundreds – feeding on the ground, once on a car park, at other times on weedy shingle, moving forward like a mobile carpet as they picked up insects. Field guides don't mention it: but it is so exceptional as perhaps to be misleading, anyway, if you put it in. And lack of space meant it was dropped. Good, or bad?

I'm often a little surprised by what I overhear in bird hides. Beginners tend to leap to conclusions based on one feature without considering others. A tufted duck, with its yellow eye, must be a goldeneye. Two people argued about a wader with a curved bill – curved beak, must be a curlew! Never mind that a curlew's beak curves down and this one curved up – it was actually a greenshank. Two others debated another wader – and they had a book with them. They looked at the pictures but didn't read the words. Look, it's brown – it must be a ruff. Or is it a snipe? It has a long beak – yes, look, it's brown, with a long beak, it's a snipe. Forget the brilliant red legs – it was a redshank. I heard about someone trying very hard and finally convincing herself that she was watching a willow warbler, even though the bird was going *chiff-chaff-chiff-chaff* all the time. People may need a lot of help, and a book is no use to them if it doesn't give it.

I'm not making fun of people. I make similar mistakes talking about things I don't know about. It is just to illustrate how basic a book must be to get to the root of the problem for people at the beginning of their birdwatching career. You learn as you go along if you grow up looking at birds from the age of five, but it can be extraordinarily difficult to grasp if you start at 55. A bit like learning a foreign language.

The simple lesson in producing a good field guide is to be clear, be unambiguous, be practical, be helpful: think about it carefully. This seems obvious, but in a world in which language doesn't so much evolve as change rapidly as if infected by a virus, it is vital. So many people now say no when they mean yes...

Writing a field guide – an identification guide – leaves little room for personal style or individuality. After all, you are putting down the facts, which have little scope for interpretation. Yet there is a little wiggle room in both text and illustrations for suggesting more about how a bird looks, or moves, or sounds, beyond its basic shapes, colours and patterns. Books do vary a bit, thankfully. Although each one is trying to show the bird exactly as it appears, you can still tell the different styles of bird-guide illustrators. Perhaps you can pick up a touch of the style of their writers, too.

There is another interesting point about writing a guide. It may not apply to luminaries such as Jonsson and Svensson, but for myself there are species that I have studied closely, others that I am very familiar with, and I have written identification texts from my own viewpoint, good or bad, right or wrong, using my own experience. Other species are just everyday birds that we know pretty well, more or less. Others have complex plumage sequences or moults that need a good deal of research to get right. And then there are those that I have never even seen. Quite obviously, there is a lot of researching (stealing, copying, whatever you like to call it) of material from elsewhere, other books, various websites. Maybe even a risk of perpetuating errors.

Bird artists can do whatever they wish and interpret what they see however they like, just as other artists do, but a field-guide illustrator must take accuracy to a different level. My old friend Peter Hayman required frequent long journeys to the Natural History Museum collection in Tring and many hours poring over references, but the results were extraordinarily accurate depictions of the birds from above, below and the side, perched and in flight. They naturally had a repetitive style about them, ideal for analysing the plumages and for comparison, and some people found them a little 'stiff' (Killian Mullarney told me he found them a touch too 'stretched'), but his work at its best became *beautiful* works of art, too, painted with great skill and sensitivity. His finest bird-guide work was immaculate and incomparable. It is so easy to dismiss such things without really appreciating the years of effort that have gone into their creation.

Roger Tory Peterson's 'patternistic' approach did not necessarily put in all the detail, but gave us the essentials of the birds so that we could work out what they were and compare each species with others shown at the same scale, at the same angle of view, in the same lighting. Yet illustrations used long before, in the ground-breaking Witherby *Handbook of British Birds*, were remarkable and full of (mostly) accurate detail; these, by artists such as the Dutch expert J. G. Keulemans (originally published elsewhere in the 1920s), were extraordinary things and still repay careful scrutiny.

Archibald Thorburn was a British artist who produced a marvellous volume of mammals and also his own bird books, in 1915–16, which were a remarkable tour de force. In 1967 they were edited into a single volume with a new text by James Fisher, doing us all a huge service and creating a beautiful book. The standard of his painting is phenomenally high given the stiff museum specimens and mounted skins that he must have worked from. To produce a complete range of species including many rarities, all in great detail and highly accurate, must have been a gigantic task.

Some more recent major artists may be in danger of being forgotten by new generations of birdwatchers. Charles Tunnicliffe, a Royal Academician, created wonderful paintings. He was also famous for his measured, life-size drawings of dead birds, brought to him in his Anglesey studio by all kinds of people who found (or simply shot) birds nearby, which he used to create beautifully composed and accurate pictures. And PG Tips cards, but not field guides. The sheer brilliance of Robert Gillmor was not really exploited in a field-guide context, although he was perfectly capable of doing some fine identification plates (see the *Herons Handbook*). But he is best remembered for other things, brilliant things. Donald Watson was not really a field-guide illustrator but contributed a lovely set of informative and attractive plates for the *Oxford Book of Birds* in 1964. Painters such as Norman Arlott, Alan Harris, Ian Willis, Ian Lewington, Dan Zetterström and Killian Mullarney have all moved the art of field-guide illustration on still further, each adding their own brand of brilliance. A female name, too, at last: Hilary Burn is a fine artist who has done great illustrative work for a series of books from RSPB guides and bird-family guides to the recent monumental *Handbook of Birds of the World.*

Although I spent most of my RSPB career working for women – all my editorial managers, several heads of departments and for a while a chief executive – I don't think a classic field guide has been written by a woman for many years. There have been some excellent artists, of course, such as Laurel Tucker who so sadly died before she could fulfil her potential, and, as I have mentioned, Hilary Burn, who has done marvellous work for many books. But the *Observer's Book of Birds* by S. Vere Benson, first published in 1937 and revised and reprinted many times in later years, became a classic that influenced innumerable birdwatchers for decades. It would be interesting to know why there were so few *women* authors of such books.

Early RSPB essay competitions, involving descriptions of the natural world, were almost invariably won by girls. But as in so many fields, there was and sadly sometimes still is a simple disregard for women. The environmental effects of organochlorines were researched by Rachel Carson, whose scientific papers and book, *Silent Spring*, revealed all we needed to know about them. But, as my old colleague Conor Jameson showed in his book and radio programme about this, she had to fight hard to study science, and her work was simply pushed aside by many scientists and politicians (and, of course, the chemical manufacturers). What does a silly woman know about such things? Who does she think she is?

Conservation itself has a better record. The RSPB was founded largely by women, influential and respected. There have been strong figures such as Phyllis Barclay-Smith, who was influential in the International Council for Bird Preservation (which morphed into BirdLife International), latterly secretary-general. She was revealing the horrors of oil pollution back in 1930. Yet she resigned from the RSPB when denied the position of secretary in 1935, because her aunt, one of the RSPB's founders, believed a man would be better for the job. She continued in other major roles and was eventually made MBE, later CBE, for her services to conservation.

There were others later, such as Janet Kear, a biologist specialising in wildfowl, who became president of the UK's senior society, the British Ornithologists' Union, and editor of its scholarly journal, *The Ibis*. The RSPB had two excellent and long-standing women editors of its magazine, *Birds*, and in less public roles was full of female heads of departments and sections, from less ornithological (such as marketing) to research-based roles. Barbara Young (now Baroness Young of Old Scone) was a popular, powerful and successful CEO. The current CEO is Beccy Speight, who moved over from a similar role at the Woodland Trust.

But while women have performed such roles in 'conservation' terms, there were traditionally few women actually going out birdwatching. Then again, how many go fishing? It has been a male activity, a white male activity, for a very long time, but fortunately things are changing. When I was on the *British Birds* Rarities Committee we tried hard to recruit a female member, but found it difficult. There are a couple, now, but the first 70 were all male. In the 1970s, though, there were one or two well-known women on the rarities scene, who were very much involved and welcomed into the whole early twitching activity with no fuss or bother at all.

Hopefully things have improved and will continue to improve, but it is still all too common to see female birdwatchers ignored by the men in hides, and of course there are still problems of homophobia and racist attitudes. This is getting too far from my theme, here, but just worth noting: there is absolutely no place for any of this.

When it came to the text, styles and approaches were also different. Richard Fitter distilled the essence of a bird into a few lines but didn't have all the forensic detail that became available to later authors as birdwatchers became more and more intense in their observation and demanding in their requirements. Later authors, including Steve Madge, James Ferguson-Lees and particularly Lars Svensson, continued to redefine and perfect the concise field-guide identification prose. Lars Svensson also did the essential 'in the hand' guide for ringers, with huge detail – which has since been complemented by BTO guides, another ringers' guide from the French author Laurent Demongin, which is remarkably comprehensive, and by various websites such as blascozumeta.com, a brilliant Spanish addition to the increasingly huge repository of knowledge about identifying, sexing and ageing birds.

Ian Wallace added his own brand of identification prose to the fantastic nine-volume *Birds of the Western Palearctic*, published from 1977. The sections on field characters allowed him enough space to amplify the basic descriptions (which were, anyway, covered by detailed museum-style plumage descriptions later in each account), and he added in a great deal of behaviour, simple movement and action – the 'feel' of the bird, how it flew, how it twitched and bobbed and walked or hopped. Some people found all of this a bit over the top, unnecessary, not very useful to the average observer, maybe a bit pretentious, while others though it was absolutely great.

The intensification of identification texts forced a division between the ordinary, everyday person with an interest in birds and the more technical identification expert who needed to know about tertial lengths and emarginations, greater-covert tips and hind-claw length. Specialisation left many people 'behind', but most were quite content to enjoy birds in their own way, every bit as valid as any other.

In this book, I have often used un-field-guide-like language. When I was still at school I had a book by Kenneth Richmond, *Birds in Britain*, published in 1962. It just has chapters about birds written in a very simple but appealing way, not in any kind of strict order or particularly systematic. I recall, for example, that he wrote 'The bittern is a queer customer, from any angle' or something similar. That would be good in a field guide... but would be edited out, not being sufficiently accurate or informative. What does it *mean*, exactly? But I *remember* it. And it is *true*! There was a lot more to be learned from a narrative approach than by simply poring over a field guide with its closely cropped language and too-tight space, but maybe few people now want to bother with it.

Reading a book also has a comfort factor. It may be compared to watching *Match of the Day* or reading a newspaper (more likely a website) report of a football match even when you know the score, maybe were even there. We all like to see something we know, to feel comfortable with it, to be reassured by it: you know it already, but it's nice to see that you were right, that other people agree with you – there's an important little bit of satisfaction and confirmation there. You are then, after all, as good as the book.

GET A NOTEBOOK AND PEN

Of course, you won't. But I'll try anyway: please read on. I am not unaware of the extraordinary benefits of a smartphone and all it brings: apps to identify birds and their voices, instant reference to the internet, whatever. Brilliant, unbelievable in so many ways. But the old way still has its uses.

Teachers helping pupils to revise before exams encourage them to *write everything down*. It goes in better that way; it *sticks*. Recent researches remind us that witnesses to a crime (or people questioned after being shown a CCTV video of one) 'remember' seeing all kinds of things later, especially if prompted by another person. This applies to all kinds of people, some you might expect to be observant enough. Women with dark hair become men with balaclavas, skin colours inevitably change, weapons are remembered where there were none, numbers of criminals go up (rather than down) and so on. And where were you last night at about 8 o'clock? No idea. How about last June?

Clearly, they may not have been expecting to be quizzed on these things at the time. This is pertinent to identifying birds: people in general aren't deliberately or consciously getting things wrong, and they *believe* their *recollections* are correct. Birdwatchers can't be much different, except that experienced ones or careful beginners are likely to be more objective, trying to analyse what they see at the time rather than later on, reducing the room for error.

Yet, given these results, maybe even the most expert observer trying to make the most meticulous, objective notes *after* seeing a bird will presumably be making mistakes from time to time, but never know it. Inexperienced and eager people, hoping to see a new bird, perhaps, are likely to be making more. There is always the risk of 'seeing' what we wish to see, rather than what is there. And that is not something to be ashamed of, as it must apply pretty much to everyone: and, as we all know, birds in all their variations, seen in all kinds of circumstances and changing light, are difficult!

My late friend Ian Langford was going to produce a book of field sketches to show how the artistic birdwatcher could use observational skills to help identify birds.

Sadly, he died before he could do it, and, of several artists who were selected and agreed to contribute, only D. I. M. (Ian) Wallace had provided a set of sketches and written accounts for the book. I recall times long ago when lesser observers criticised Ian, saying his sketches changed as he watched a bird and an ordinary chiffchaff or reed bunting turned into something rarer, but these people failed to appreciate the process. He was looking, re-evaluating, looking again, learning more, seeing more, realising he had one bit wrong and putting it right, until the bird gave better views and all came clear and his identification process reached a proper conclusion – common or rare. The value of looking and looking and *looking again* is immense, and making notes and drawings forces you to do it in a disciplined way.

Many times, I have tried to sketch something or note down details in a notebook and, between looking at the bird and taking up my pen, have forgotten a detail: was it six white spots on seven primary tips, or five white spots? If the bird was still there, I had to look again. This process in itself is useful, forcing a second look and maybe a reappraisal. But if the bird has gone, the truth may have disappeared along with it – who will ever know? It may not matter, but sometimes, it will.

It remains essential to carry a notebook and pencil or pen (hopefully one that works, even in damp weather) and to use it. Writing anything down helps reinforce it in our memory anyway. Trying to sketch it, however badly or basically, does the same and also has another benefit. You cannot truthfully draw the shape of a beak or tail unless you *look* at it; and you cannot draw the beak and put a label on it such as 'black, yellowish at base of lower mandible' unless you *look and see.* If you want to label the leg colour, you have to look at the legs: if you've done the sketch and can't write the colour, you have either forgotten it, or didn't look, so must check it again.

This sounds obvious but it really is vital. Once when I found a ring-billed gull (it was the first for Britain) I sketched what I could see on it, but because there were no useful books to hand – and certainly no internet – it was only later, after a friend phoned an expert (Peter Grant), that I found that the *eye colour* is crucial. I looked at my sketches and saw that I had drawn a pale eye. But I had not *labelled* it, and I couldn't *remember* it. Was it just a natural reaction to draw a pale eye on a gull, or did I really see it? Too late now, it was already getting dark. Fortunately, the bird stayed around and the pale eye was seen well by many people. How well I remember an older, more experienced birdwatcher – A. J. L. (Tony) Smith – saying 'Tick off pale eye!' as he watched it a few days later. But I should have looked at my notes and sketches, realised I hadn't got the eye colour, and looked back at the bird that very first time. It could have got away.

More trouble comes if you try to make notes and sketches *after* seeing the bird, once it has left the scene: there is no hope of a second look. There may be no choice if the view is brief. But these troubles are multiplied if you look at a book first, before making your notes. Then we get the 'imaginary evidence' coming in – the bird will have been carrying a knife or a gun, wore a black cap and a mask, had at least one accomplice – like the criminal evidence videos. In other words, if you read that something should have a black eye-stripe, or a pale wing-bar, or a white spot on the tail, it suddenly becomes all too very easy to 'remember' these features, whether they were there or not.

'Pale wing-bar? Oh, yes, I'm sure I remember seeing that. Four spots? I saw spots – was it three or was it four? – it must have been four.' The book says four, so your mind switches to four, even if there were only two (or perhaps five). Having read all the important features in a book *before* setting about scribbling your notes really does make it next to impossible to get the true features of the bird you watched onto paper.

But if you don't do it at all, and try to remember a few days later, let alone a few weeks, you have no chance.

The only solution, if the bird hangs around long enough, is to make notes, and sketches if you can, *while watching it.* If it flies off too quickly – a passing seabird maybe – do so immediately afterwards. If you are with other people, it is a good idea to shout out features – 'It's got a wing-bar', 'Can you see the dusky underwing?' – but this does carry the risk of one dominant observer planting (albeit entirely unwittingly) his or her views into the minds of others. 'Erm, well, no I can't really see it, but maybe you're right. You're better than me.'

If you are drawing the bird, obviously do the best you can but it is not necessary to try to create an artistic masterpiece. Indeed, a more basic, simple sketch with labels can be better and more honest, without the attempt to produce some sort of perfect image. There is a problem, though: you might quite legitimately see 'three or four' marks and could justifiably write down 'three or four dark spots', but you can't (unless you are some sort of genius) *draw* 'three or four' – if you just didn't see enough detail, a sketch might try to make you force the issue. Beware, that's all.

If you are photographing the bird, all well and good, but the camera will only 'see' in a certain way, catching tiny moments of time, and results might not be ideal. And pressing a button does not impress details into your mind in any way, whereas looking, seeing, asking questions and writing down what you see (but not what you don't, except to note the absence of something expected) is the way towards total recall later on.

Either way, the important thing is to ***carry a notebook and to use it.***

Use a notebook outdoors, while you are birdwatching. Most people don't bother. I would recommend it as a necessity if you want to make progress. Take notes; tot up numbers; make little drawings, however basic. Look hard at birds and note down what you see: it is by far the best way to learn and to work out difficult identifications. As you progress, you won't need to do this so often, but if ever you come across something difficult, or rare, keeping a written and sketched record of what you see is vital. It makes you look and look again, to check what you think you have seen and to put it down, so you don't forget. And it helps convince other people, too.

Obviously, you will not be sketching things very often, and eventually you may only sketch birds that are unusual in some way, maybe rare, maybe in a strange plumage – nobody suggests you carry on taking descriptive notes on ordinary familiar birds.

My little notebooks carried about in a pocket have long since disappeared, but I used to make copies of sketches and notes with my records in a larger hardback book, and I still have many of those. I never felt a list did justice to a good day, a good place or a good bird... they needed more, something more like a little diary entry. Many items in this book are lifted direct or expanded from these notes. I took great care to be sure that the drawings reflected what I had sketched, but they were usually just drawn in to a ruled exercise book and only occasionally worked up into more finished drawings or paintings on a decent bit of paper.

My descriptions vary greatly according to the bird in question. I always took a lot of notes on gulls. I would write about, rather than necessarily sketch, such great birds as yellow-browed and Pallas's warblers (coming from Siberia, after all), which never fail to get the adrenaline moving. So my books vary from lists with no notes, to lists with short or long written notes added, to the same thing with drawings as well. Here are a few examples. They are not intended to be great or ideal examples of their kind, nor are the drawings meant to be perfect, which would be beyond me, but they just show the sort of thing you can do.

You are highly unlikely ever to make such sketches and notes on a common bird such as a blue tit, but why not, for example, practise a bit on a reed bunting or meadow pipit, or something similar with nicely detailed feather patterns? Or try to do what an artist does (if doing things properly) and draw what you *can see* and

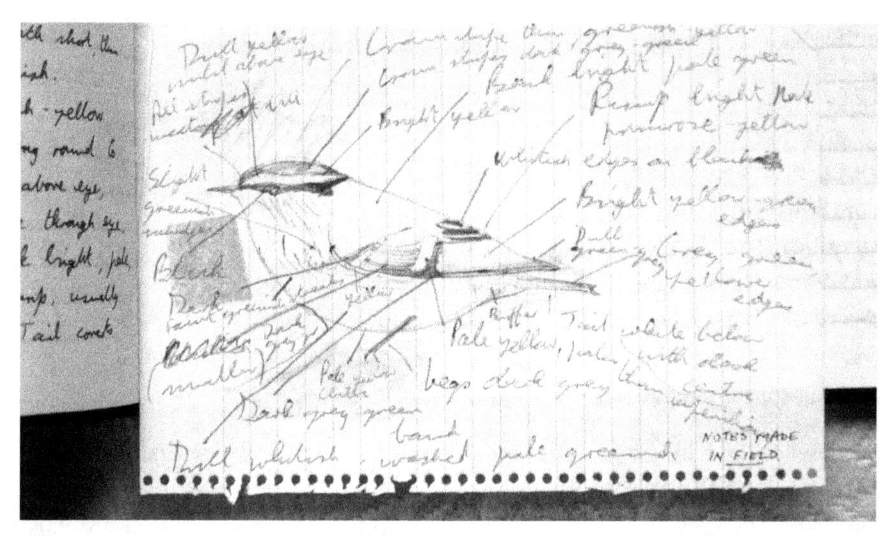

***Pallas's warbler** is no longer a major rarity, but it's such a good bird that it interests and excites me and I can't resist giving it the full treatment. Most of my grubby sketches done while watching the bird have gone, but here is one, with the tidied-up version later transposed to my more permanent book. It does nothing for the bird, which is beautifully delicate and colourful, but shows the distinctive features well enough. This warbler is about the size of a goldcrest, and has flown all the way from eastern Asia.*

Here's another scrap of paper saved from a little ring-bound notebook used long ago, a ***black-throated thrush****. Your notes and sketches do not have to be neat and expertly drawn, so long as they are done on the spot. The dark patch at the top is where it was stuck inside a bigger book and the Sellotape glue shows through. It was made originally in black ballpoint pen, not bad so long as it doesn't get wet. This is another vagrant from Asia, about the size and shape of a blackbird. In spring the male's gorget firms up into a solid patch of matt black.*

And another, watched in Essex: a ***Naumann's thrush****.*

no more: so, if the bird is a bit distant, how do you cope with what you know is a streaky back when you can't see the streaks? It is difficult: you can say or write 'a few streaks', but do you draw three, or five, or six? Or what? Or do you strain to see exactly how many you can actually see, so your verbal description becomes more precise, the drawing more honest?

I always felt a little proud of my drawings of the olive-backed pipit. A version appeared in the Scilly bird report and was seen and later referred to by Peter 'PJC' Conder, the excellent past director of the RSPB (before that post was referred to as CEO). Peter was another person whom I met a few times but hardly got to know. He was still there when I first went to the RSPB headquarters in Sandy to discuss gainful employment, and he was always good enough to speak to me on his occasional visits, usually to look up something in the library, long after he retired. When I was editing *Birds*, I sometimes tried to encourage whichever CEO was writing the introductory piece at the time to put a bit more emotion into it. Graham Wynne would be apoplectic about some development but write about it in a very considered, professional way. I wanted a bit more oomph. Of course he was absolutely right. But in those magazines that I read as a young member, I remembered Peter Conder being forceful and persuasive, more emotional, and it hit the spot while not being any less than entirely professional and responsible. He was a bird observatory warden on Skokholm in Wales for a time, and people

*This **olive-backed pipit** was based on field sketches made on Tresco in the wonderful Isles of Scilly, which I have not visited in many years but remember so well. The bird itself was found by my long-time RSPB colleague, Chris Harbard.*

now perhaps forget his interest in bird identification. He pointed out that an olive-backed pipit had a pale ear-covert spot, like a rearward extension of the superciliary, cut through by the thin black eye-stripe. It is sometimes echoed by a tree pipit but not nearly so clearly, so it is a useful feature, and PJC remarked on this in a letter to *British Birds* and referred to my drawing, which showed it well. I did not know about it as a particular feature at the time, but by looking hard and drawing what I saw, I got it quite nicely in the sketch. Just look, keep an open mind, don't worry about what the books tell you that you should be seeing, just note what you are *actually* seeing at the time.

Here are notes on two **rustic buntings**, *one in Norfolk, one unexpectedly in Cambridgeshire: I didn't find either of them, I just went along to look. They have pages of detailed written descriptions, too! Not that I am likely to read them in full again.*

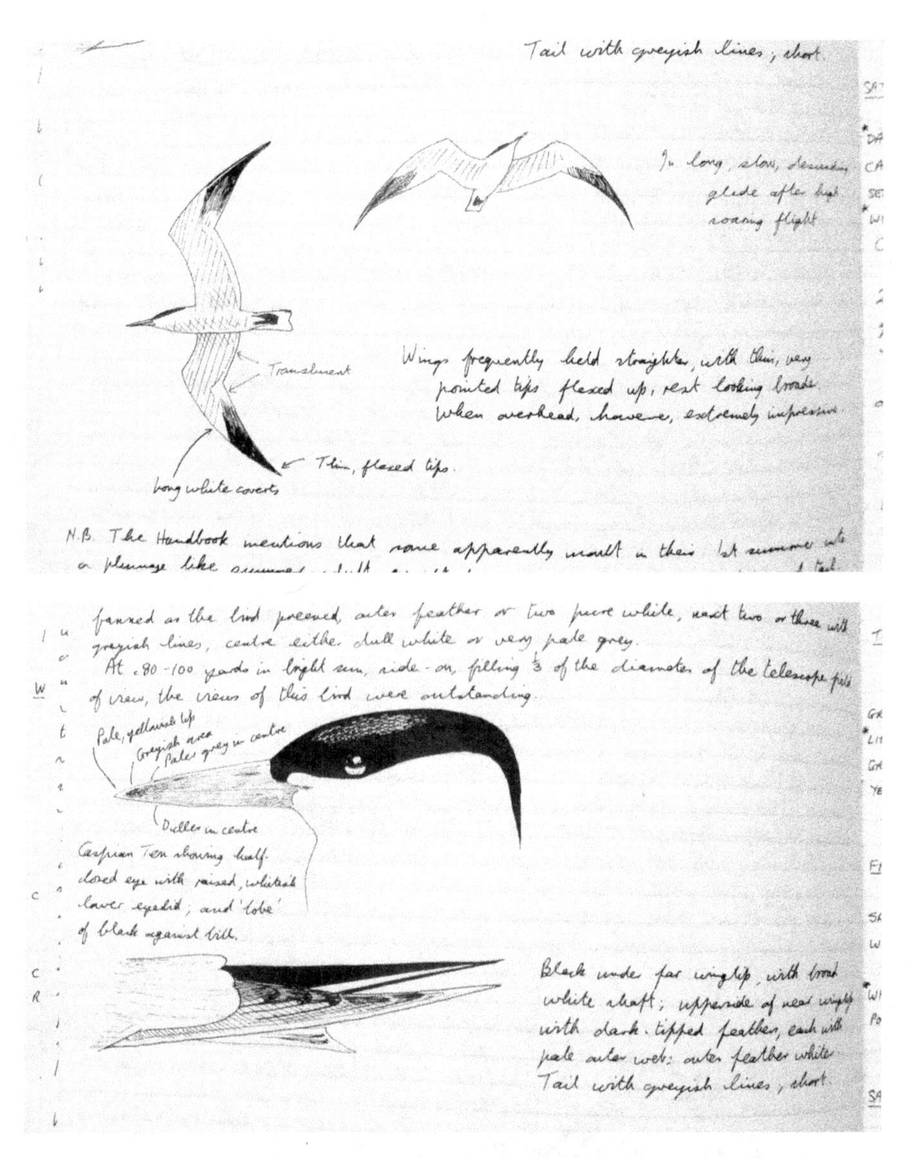

Not a gull, but nearly! Sketches of a **Caspian tern**. *Earlier sketches of it were more basic, less detailed and lacked such small things as the 'lobe' of black that turned down against the base of the bill. Going back, looking again, trying harder, always improved understanding and added to the accuracy of my descriptions.*

The black patches under the wingtip can give the angular-winged Caspian tern a bit of a mini-gannet effect. The bill is huge and looks it in typical close-up illustrations, but in reality, on a bird at moderate range, it is entirely proportionate and does not look at all out of place. Indeed, big and powerful it may be, but it is an elegant bird altogether.

It is, of course, simple enough (and quite understandable) to say 'Caspian tern!' and leave it at that. Up to you. No rules apply.

I later amended my notes on the rustic buntings by saying, for example (on the first, pencil, drawing) 'Head smaller than this' – the drawing began to look like a Lapland bunting whereas the bird really looked more like a reed bunting. With so much detail I had made the head bigger to accommodate it. And on the second drawing I noted that I'd missed something from my field sketch, even though it seems to be on the drawing made on an earlier visit. And did the superciliaries meet over the bill or not? From the side I didn't need to know to make the sketch, but trying to do a 'top of the head' view, I was forced to decide... and didn't really know. Such a detail may be of no importance, but you never know. Also, the wingtip and vent were left blank in a final sketch: not noted in the field, so they couldn't be filled in.

Of course, a photograph – the more likely option nowadays – would have filled in the blanks and made some things more accurate, but would it have helped me understand the bird as well as having to look, look and look again, to make these notes and drawings? The answer is obvious: *do both*! A good photograph is a different but hugely satisfying achievement, but you may not remember the details 10 years later.

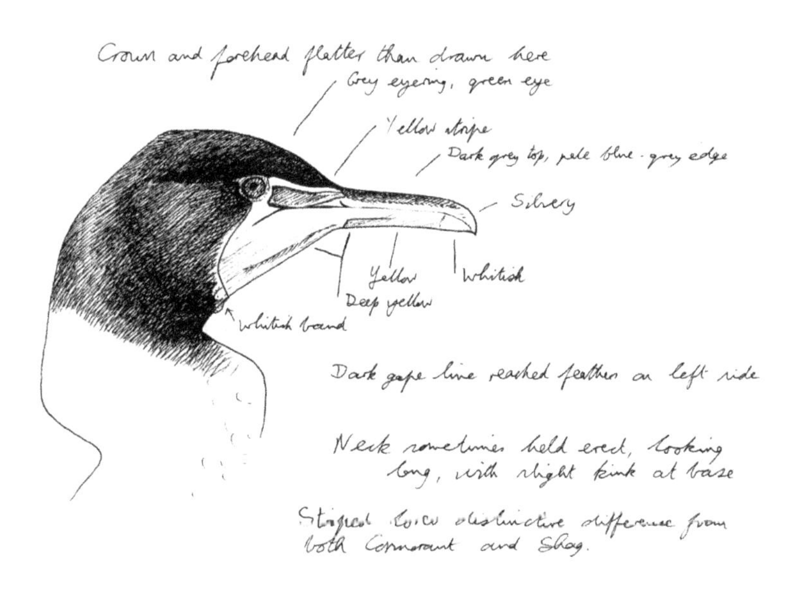

This is a bird I travelled a long way to see. I wonder, if I had walked past it without knowing, would I have given it a second look? It was thought to be both cormorant and shag before being properly identified by T. J. Williams, to his great credit: a **double-crested cormorant**, *all the way from America.*

There is little that might be confused with a jackdaw... but that doesn't mean it is not worth looking at. Jackdaws seem to enjoy life, keeping up with the rooks and often producing an extraordinary cacophony of calls.

Sorry, this is turning into a pile of rarities, which is not at all what the book should be about. More ordinary things will follow, I promise.

Indeed, a statement that should be writ very large is **look at common birds too**! After all, 99% of the birds you see will be common and you may as well enjoy them to the full. But I have known people with a high reputation in rarity circles who hardly looked at common birds and just couldn't remember some very basic features, which had a big knock-on effect when it came to comparing rarities with potential confusion species. To identify and also to really appreciate rare birds, you need to know your common stuff, preferably inside out.

DISTRIBUTION, HABITATS AND TIME OF YEAR

No-one *needs* a really technical understanding of habitats and ecosystems to see birds, but knowing roughly where to find them, in terms of the kind of places they prefer as well as the geographical area they occupy (distribution) is a huge help.

Distribution has been dealt with nicely by a series of studies and surveys. John Parslow (who once interviewed me for a job) long ago did a study of local bird reports and wrote an important series of papers on the status of birds in Britain. Then came the *Atlas of Breeding Birds in Britain and Ireland* (1968–72, published by Poyser in 1976), followed by a *Winter Atlas* (1986), then the *New Atlas* (1988–91, published 1993), and then another covering the years 2007–11 (*Bird Atlas 2007–11: the Breeding and Wintering Birds of Britain and Ireland*, published 2013). It was pleasing for us Midlanders to see the pioneering work of C. A. Norris in the West Midlands in 1950 followed up by an atlas based on 10 km squares produced by the West Midland Bird Club (by John Lord and Denis Munns) in 1970, a project to which I was proud to contribute a tiny part. Many people used to focus on a local 10 km square or two, while having a spare set of 'atlas cards' to record breeding species found elsewhere, such as while on holiday in more remote areas. We all recorded presence, possible breeding, probable breeding and proved breeding according to a set of agreed criteria. It's singing! Probable breeding, then. Look, it's carrying food! It must have chicks nearby. Now it's carrying a faecal sac (a dropping) – it *has* got chicks nearby. Proof! Many individual species studies have reflected recent changes in numbers and distribution.

Habitats are fairly easily classified in a standardised way, but for birdwatchers a basic idea suffices to suggest what bird species might be expected, in the right place and at the right time of year. The BTO uses 10 habitat types that are further subdivided at three levels. The basic 10 are: woodland, scrubland, semi-natural grassland and marsh, heathland and bogs, farmland, human sites, water-bodies (freshwater), coastal, inland rock and miscellaneous.

Timing can be crucial: classic cases would be a wagtail showing a lot of yellow in winter (which would therefore be a grey wagtail, not a yellow wagtail) or a small, streaky, pale-legged pipit in winter (which must therefore be a meadow pipit, not a summer-visiting tree pipit or a dark-legged rock pipit). Combine a map with a calendar and you can see, for instance, that you might get ring ouzels on the south coast in March or October, but in summer they would be farther north or west in breeding areas; or you might see migrant dunlins in the south in spring and autumn, as well as large numbers spending the winter, while in summer they're gone, or breeding up on the hills of Wales and Scotland.

Habitats include such categories as arable farmland (wheat, barley, rape, root crops and the like), pastoral land (with sheep, cattle and so on – maybe llamas or alpacas these days), with or without hedgerows or shelter belts. Much depends on the underlying rocks and soils, whether there are acid grits and sandstones or limy chalk and limestone. The rocks may play a greater part in affecting the vegetation and perhaps the butterflies than the birds, in most cases.

A wagtail with a lot of yellow on it, on a stone ledge beside a little spring with a 'travellers' cup' on a chain: this is a grey wagtail. A yellow wagtail would be highly unlikely in such a place. Anyway, this one has short pale legs. Habitat, time of year, physical features, all come together to identify it.

Farmland has notoriously lost much of its hedgerows, and many of the remainder are cut hard and often, leaving either few blossoms or few or no fruits (pity the poor dormouse), or insects, and a loss of likely nesting places for many birds. More insidious are the pesticides. In the 1960s, and into the 1970s before DDT was banned, birdwatchers were very aware of the savage effects of agricultural chemicals. This was the era of persistent organochlorines (aldrin, dieldrin and so on) that killed huge numbers of small birds. Birds of prey ate the contaminated birds and the pesticides concentrated in them – and they became infertile, laid eggs with shells so thin that they could not safely be incubated, or just died. Much of the countryside was thus devoid of birds of prey. Since the control of these pesticides, things improved, but now we face problems with new ones that just wipe out everything in the nature of a pest as well as nearly everything else as well. 'Weeds' are long gone, and insects, mice, voles and the like are hard to come by, so we get vast areas of farmland almost devoid of wildlife. It is a huge dilemma as there are so many people to feed, but pesticides affect us too. And you may notice an absence of moths, butterflies, even old enemies such as greenfly and blackfly, from your garden, even if you don't use a pesticide from one year to the next yourself.

It is, I believe, important for younger birdwatchers now to realise what things were like just 50 or 60 years ago. In the DDT era people could see the bird populations declining fast; then came problems such as the development of winter wheat, so that, instead of stubbles or ploughed ground left through the winter, followed by short crops in spring, we had dense, tall crops by midwinter. This meant that such species as lapwings and skylarks, which need a patchwork of habitats and some bare earth or short vegetation in which to feed (and in which their chicks can move about), simply disappeared from large swathes of lowland farmland. The annual displays of courting lapwings, tumbling and calling over the fields, became no more than a memory. The delight of lying on the grass and listening to the sparkling songs of larks overhead has long gone from most places. We 'get used' to what we have (the 'shifting baseline syndrome'). In my infancy and early school days I watched pylons going up across the countryside (and huge pipelines being laid for gas, which did not leave the same visual legacy): these have long been so 'normal' that nobody now will really think of how much they scarred the landscape or ruined the irreplaceable view. Perhaps windfarms and solar farms will become accepted in the same way, although with greater consequences for wildlife.

Many people today don't know any different, but birdlife in farmland years ago was so very much richer. Even in mid-Wales in the 1970s I was noticing the expansion of close-cropped uniform pasture full of sheep at the expense of the 'fridd' zone on the hillsides, with its bracken and heather and short grass full of larks, wheatears, whinchats and meadow pipits, so rich in colours that varied with the weather and the season, so much a part of the landscapes of Wales. Much remains, but so very much has been lost.

Farmland merges into other open-ground habitats, such as downland. Old-fashioned flower-rich downs are really only left in nature reserves, which reveal just how wonderful such places are (but even these are less than they were). There are heaths, too, with more gorse, heather, bracken and scrubby bushes, which in turn merge into woodland. This is where you might look for linnets, stonechats, tree pipits, nightjars... Higher up, generally, are the moors, but moorland comes down low in the west and north so is hardly really an 'upland' habitat although that is a convenient shorthand term for such places. There will be grassy areas, bracken slopes, heather (short and long), perhaps eroded peat giving little peaty gullies and miniature cliffs, as well as rocky outcrops, deep ravines and fast-flowing streams, perhaps peat bogs with stagnant pools. In the north and west, peat bogs become more extensive and create extremely important habitats. Drier areas may have wheatears and whinchats, ring ouzels and birds of prey, boggier places perhaps dunlins, or snipe, even greenshanks up north, and streams are good for common sandpipers, dippers, grey wagtails.

Rocks and cliffs add extra potential for such birds as ravens, peregrines, even eagles, while higher plateaux can have breeding dotterels and ptarmigan. The Scottish coast still has native rock doves on a few cliffs (elsewhere the doves are domestic pigeons 'gone wild'); in Wales and Ireland, the Isle of Man and a very few southwest Scottish locations, there may be choughs. Brilliantly, they have also returned to Cornwall, where they have ancient connections.

In addition to all the natural and semi-natural habitats there are many kinds of artificial ones and 'edge' or marginal land. As I describe elsewhere, many of my early wildlife discoveries were in an area of mining spoil mixed with natural heath and bog. Fantastic it was, too! But while many species can find a home in these areas, in places such as the ever-increasing retail and industrial parks most simply cannot survive, while a few thrive. The pied wagtail is a typical car-park/tarmac/concrete specialist. There are vast areas of towns, cities and suburban housing, and especially roads and all their associated roundabouts and slip roads and this and that, where few birds can be expected. Even those that might survive will only feed there and can hardly nest in such a place. There is a bit of 'you get what you deserve' about these areas.

And sadly, many 'green spaces' in otherwise more or less urban areas are lauded (rightly enough) for their benefits to people, but when they are also cited as 'good for wildlife', the evidence is usually slight or non-existent and the people who say so seem to have little idea of what they are on about. In many 'green spaces' you couldn't seriously expect so much as a magpie or a house sparrow. Birds need food (and time to eat it undisturbed), secure, safe resting and roosting places, somewhere to nest where there is not constant danger of being scared off the eggs or leaving chicks exposed. A field of short grass covered in dog walkers and kite fliers will not usually cut it.

There are better things, of course. Who would have thought we might watch little egrets as a matter of course and, now, in places, even groups of great white egrets? Kites overhead, more buzzards, the chance of a peregrine. But never a turtle dove purring gently from a blossom-filled hedgerow, no more tree sparrows around the old, decaying oaks. No more song of the curlews in green Midland river valleys, or the thrum of the lapwing's broad wings as he throws himself excitedly around the sky in spring. No wonder the 'old names' – country names for many of these birds – are long forgotten: there is no longer the same close link between people and the birds as before. I used to be able to lie in bed and hear a cuckoo singing... I could, sometimes, on a quiet night, hear the sudden crescendo of a distant steam locomotive marshalling its coal wagons and momentarily losing footing on the rails. The steam locomotives have long gone, and so have those cuckoos.

None of this, however, should be interpreted too rigidly. Listing birds in each of these habitat categories is apt to mislead. They fly about! Look, for example, at a bit of coast. A little estuary with some saltmarsh, shingle, maybe a sandy or muddy creek, shallow sea running out to a bit of stony mussel scarp. You will probably see the standard estuary birds, depending on the geographical location and the date. So, perhaps gulls and terns, a redshank or two, oystercatchers, maybe an eider, a rock pipit. No terns in winter, but there could be more waders. But look at a bit of beach with some groynes retaining sand or shingle, a little earth/sand/grass-topped cliff. Here the typical beach birds could be carrion crow, rook, feral pigeon, woodpigeon (oh yes, they do), starling and pied wagtail. In spring perhaps a wheatear on the groynes, even an adventurous wandering local robin or stonechat, sand martins in the earth cliff. Not much 'estuarine' or seabird about them. Birds simply go where they can find food – and waves gently washing the beach will fetch up all kinds of scraps, including insects and seeds, hence the pigeons.

Sit in a hide by a lake, overlooking a bit of wet gravel with a few puddles and the edge of a reedbed, and a few overhanging willows, and you might see a stock dove coming to drink, a pheasant marching through the reeds, a bullfinch and a great tit in the bushes, a swift overhead, a wren bobbing about beside a moorhen, a cuckoo looking for nests. Look up 'wetland birds' in your book and none but the moorhen will be there. Several times I have listed 'pheasant' as a reedbed bird, only to have it cut out by people who know better: but if you go to a dryish reedbed or look at the bushy fringe, you are very likely to see a pheasant. Keep that in, then.

You will get to know how things go... but all the same, habitats are crucial to finding your birds. With a bit of common sense it is not too difficult to see 'what should be there' and to understand that a passing gull is not actually a garden bird or a woodland bird, but just happens to be flying over. It might still be a good bird for your 'garden list'. But garden birds proper, in and around the garden, include a smallish number of mostly obvious species, and these should get you started.

COLOUR AND LIGHT

This is something that fascinates me but bemuses me completely, at the same time. I hesitate to write about it, as it will reveal the depths of my ignorance, like an infant trying to understand a PhD thesis. But birds and colour and light are inextricably linked in my mind, as the effects of light on birds – individuals or flocks in an abundance of different settings – largely inform the way I appreciate them.

The subject of colour in its simplest form, without straining your brain to think about the details of light itself, and how that might work, is well worth considering. Really, when looking at birds, it is everything. Look at the sheen of purple and bronze on a lapwing's wing, the astonishing blue of a peacock, the subtle orange-reds on a robin that glow from the depths of a holly bush yet can seem remarkably dull in a clear view at times, the clearest pearl-grey on a black-headed gull.

While writing this I looked out to see the autumn sun sinking through a glow of gold that infused the pale blue sky in the west. I had to nip outside to look, and there, to the east, the sky was still richer powder-blue with grey scuffs of cloud, behind a belt of tall leafless trees that were brilliantly pale in the low light and, through and behind them, a series of vivid white dots rising and falling with a particular rhythm that spelled out gulls, heading to roost, tilting up to the wind and gliding down again, as they do – beautiful. Indescribable really, despite my poor effort to give an idea of the picture.

It was just a moment, common and ordinary enough, yet in its way quite dramatic and splendid. The birds were just an incidental part of it, but added that little extra movement, allowing me to identify with something familiar doing something I see so often... part of a colourful scene that lasted just seconds before the light changed, the colour became altogether different, the white gulls became dark spots against the sky. Things change so much, so quickly.

Close up, looking at the birds themselves rather than as part of the landscape, they are equally subject to momentary effects. I've watched gulls an awful lot. They are mostly grey, white and black. They can look orange against blue water, or blue against an orange sunset, or dull and drab on shiny water. Against a dark sky they look white, against a light cloud they look nearly black. In the flat light of a snowstorm every last nuance of tone is revealed to perfection.

In a recent book I wrote that the blue tit can look pale and nondescript against a pale sky. Like much of what I write, it was cut out: any bird can look colourless on a drab day. Well, yes, but the point I was trying to make was that a close-up photo (or painting) of a blue tit in a field guide shows strong blues, greens, yellows, black and white stripes. It looks as if it should be boldly patterned, strongly coloured, with all these bits and pieces obvious to the eye. But it *can* appear almost plain, in certain circumstances. It is just always worth bearing

Not surprisingly, any bird, such as this ***curlew****, which shows a lot of white beneath, will look dark against a bright sky; yet the brown upperside will look very pale against dark heather. If you know what you are looking at this is no problem; if not, it might be misleading. Surprisingly, several people have insisted to me that, for example, 'that's not an egret, it's a heron'. With a white bird silhouetted against a white cloud, it might not be immediately obvious.*

in mind that a field guide picture is perhaps an average or better-than-average impression, but the reality might fall short (or go well beyond).

Any bird in a sunlit, leafy bush or tree might take on a bit of green. I've seen turquoise house sparrows flying over a swimming pool. Our brains make good sense of this most of the time, but now and then we can be thrown.

Birds are remarkable for their colouration. What, though, is it all for?

An early job of mine with the RSPB was to introduce film shows, including a film called *Speckle and Hide*, narrated by one of the classic, distinctive voices of wildlife radio, Derek Jones, whose repeats on Radio 4 Extra are still worth a listen. It showed the various ways that birds used camouflage to reduce the chances of being seen by a predator, but also how predators themselves used camouflage to get within striking distance of their prey.

There are obvious examples of 'cryptic' plumage, such as the extraordinary dead-leaf pattern of a woodcock, or the dead-wood effect of a nightjar – both are very hard to pick out if they keep still (with their giveaway eyes half-closed) on the right background. Bitterns are fantastic at staying hidden.

Not all camouflage, however, relies on imitation of the immediate surroundings. Many plumage patterns serve to break up an outline, or to reduce tell-tale effects of light and shade.

How many birds are darker above than below? This countershading 'flattens' the shape when lit, as usual, from above, so the dark back (brightly lit) and pale underside (in shade) become of equal tone and the 3D effect is lost – but the dark shadow on the ground beneath is out of the bird's control. Pale desert birds in strong sun may be detected only by a spot of moving darkness underneath them. Many small birds (thrushes, pipits etc.) have a pale belly and slightly darker flanks, the dark flank band just where the light catches it, again to help flatten the bird out and make it less obvious at a glance. A stone-curlew can 'disappear' against a pale background if it keeps still. Others have surprisingly striking dark and light patterns – such as the ringed plover – that serve to disrupt the outline, again making a give-away bird shape less obvious to the eye of a predator scanning the scene. A great spotted woodpecker amongst dark branches against a pale sky is a good example.

This disruption effect was used by the great all-round naturalist, conservationist, Olympic medallist, glider pilot, artist and all-round achiever, Peter Scott, who designed dazzle' camouflage for ships in the Second World War – bands and slashes of greys, blues and greens that looked weird close up but melted away at long range against the waves. Oystercatchers are sharp black and white, but a flock in flight, although obvious enough with flashing rumps and wing-bars, can confuse a predator so that it fails to make a successful strike.

There were a couple of inspiring papers on seabird patterns by K. E. L. Simmons in the journal *British Birds*. Bottom-fishing shags are dark so they can flush out fish from hiding places in rocks and weed; gannets are bright white, so they can advertise the whereabouts of a shoal of fish to other gannets a mile or more away, and also reduce their own visibility to the fish they hunt. Many gulls and terns have a white frontal aspect – foreheads, the leading edges of the wings – so they are less visible from beneath as they plunge-dive. Now, Ken Simmons was a super-accurate, expert observer of birds who knew infinitely more than I ever will, so I will never take issue with him. But I do remember very well being at Stafford College of Art and painting a monochrome picture of a flock of terns overhead against the sky. I thought I would be clever and just paint in the dark parts, giving a more active, flickering effect. The tutor looked at it and said 'I'm not sure the tones of the birds would match the tones of the sky just like that.' No, they wouldn't: even the white parts would look dark against a bright sky, or white against a dark sky. The birds would more or less stand out, whatever. Would these white leading edges and foreheads really work, making it difficult for the fish to see the birds as they hovered above? Enough to be selected for by evolution over millions of years?

Gulls coming in to roost at a reservoir or lake often 'whiffle' from side to side – like descending flocks of geese – so their bright white areas catch the setting sun and show up the location of a good, safe roost to incoming gulls still a great way off. Some of my most enjoyable moments birdwatching have been watching great flocks of brilliant-white gulls coming in against low, dark clouds, even through intense late-afternoon rainbows – a combination of birds, striking effects and vivid colour that is a rare treat.

Colour is clearly not always to do with camouflage. A territory-holding adult mute swan is sufficiently predator-proof to be able to 'dominate' the riverside scene over great distances, other swans having no excuse if they trespass. (So why are black swans black?) There's always more than one way, too, to achieve what you want – hen harriers hunt low over heather moors, for example, but the pale grey and white adult male seems to be as successful as the more cryptic brown and cream female and immatures.

Colour is often about communication – wanting to be seen, and especially wanting to be recognised, as a fine, fit, adult male, for instance. Black-headed gulls have dark hoods in the breeding season, and use them in display – but the hood tilts forwards, leaving a white nape, which can be turned towards another bird in appeasement. But this clearly isn't needed by the Mediterranean gull, which has a larger hood – another example in nature of 'there's more than one way to skin a cat'. But both species lose the hood from the face backwards in summer, and gain it from the back towards the front in spring, to reduce the period of time when they are most 'aggressive'. Then again, I have watched black-headed gulls displaying – wagging their heads, bowing and showing off – in white-headed winter plumage, too, so our simple interpretations of what is going on are probably just good approximations, with frequent exceptions.

Male red-breasted mergansers use fabulous striking colours and patterns combined with bizarre posturing to impress females. So do goldeneyes. In fact, most ducks are at their 'best' plumage-wise during winter, when pairs are formed or reinforced, a complication for a field-guide writer trying to use simple phrases such as 'summer plumage' or 'breeding plumage' – even the great *Birds of the Western Palearctic* had to say 'Breeding – does not nest in this plumage' for the ducks. But most birds have a 'summer plumage' in spring and summer, that equates to 'breeding' (or more accurately, display) plumage, when the males are trying hard to get a mate and to dominate other males.

Colour is often combined with a peculiar shape. Great crested grebes (and their rarer relatives) grow eye-catching ruffs ('tippets') and crests, with bright colours and jet black (a bold statement in itself) that double the effect. Mandarins have their remarkable upright 'orange peel' feathers. Colour can also change on the 'hard' or bare parts of a bird – the bill of a grey heron becomes bright orange or pink in spring, the facial skin of little egrets can take on surprisingly vivid hues including lilac and magenta, nondescript horn-coloured bills of hawfinches and chaffinches turn blue in spring.

Such colours, just like the display postures themselves, entail a degree of risk – showing off, making yourself obvious, looking bigger, brighter and fitter than your rivals, is the very opposite of the desire to stay safe, hide away, not be noticed – the purpose behind camouflage. We can be grateful, however, that the onset of breeding plumage produces some of the most beautiful effects. What better, more refined, more exquisite pattern is there than a summer black-throated diver? What is more perfect than a spring male yellow wagtail? W. H. Hudson, 100-odd years ago, wrote that, when you see your first male wheatear of the spring, you think it the most beautiful bird of all. Combine colour, pattern, the prospect of spring – he was probably right.

Most individuals of most species will look remarkably alike in their colouration. Thousands of black-headed gulls lined up in a reservoir roost might not include a single one that looks even a tiny bit darker or paler than all the others (I have seen a very few paler ones but nothing obviously darker). This constancy of colour is always worth pondering. Non-birdwatchers, I suspect, are not really aware of it. Birdwatchers *are* aware of it, even if we cannot always explain it. Why, for example, does a redshank have red legs but a greenshank green legs – why does a yellowhammer have a chestnut rump, a cirl bunting an olive one – why is a song thrush just *that* shade of brown, while a mistle thrush is *this* one?

Years ago, I used to see roosting oystercatchers, gulls, curlews and bar-tailed godwits on a playing field near a busy road. I was asked why the white birds didn't mix with the black ones. The lady who was asking, intelligent though she was, clearly had no real idea of what 'species' are and thought birds were all 'the same', except they came in different colours, just like cats, dogs or cows. It seemed illogical, to her, that they didn't all just interbreed and produce piebald, or grey, ones.

Of course, birds are 'species', not 'breeds', and amongst many attributes that separate them are their colours and precise patterns. My local gull roost has pale ones (black-headed), slightly darker ones (herring), darker ones again (yellow-legged and common) and some that are still darker (lesser black-back). Some lesser black-backed gulls are closer to black above, these being northwest European birds of the race *intermedius*, rather than our local race *graellsii*, illustrating the differences that can exist between subspecies. All well and good: we can identify things by their upperpart tone, bearing in mind all the variations according to angle of view and lighting conditions. A ring-billed gull stands out in a flock of closely similar common gulls because it is paler; in a flock of less similar black-headed gulls (as tends to be the case where I sometimes see them in Hampshire) they are more difficult, because their pale upperpart tones are so much closer.

The constancy of colour within species is usually quite remarkable. The same goes for most wild flowers – the precise colour of rosebay willowherb, the *exact* yellow of a marsh marigold or corn buttercup, or the distinctive blues of various speedwells, for example, repeated billions of times over.

But then there are herring gulls! These seem to have rules all of their own. We get the pale British-breeding race *argenteus* and darker ones in winter of the northern race *argentatus*. But some seem to be in between. Sometimes I see *strikingly* pale adults. And young herring gulls can have feathers that are streaked, spotted, barred or marked with anchor-shapes – enough to drive most of us to distraction. What is it with these things? Herring gulls seem determined to be as awkward as possible.

Nevertheless, a great field-guide artist such as Killian Mullarney, Alan Harris or Peter Hayman, painting a bird for an identification guide – not necessarily for a beautiful painting to go on a wall – will get the colour as accurate as possible to match a museum specimen or live bird, and it will, for most species, then be correct for millions of them. It is 'obvious', and a fact that we all 'grow up with' in our birdwatching careers, but still seems amazing to me.

As with most things, however, rules such as the constancy of colours within species are frequently broken. Various pigments introduce colours to feathers – melanin, for example, creates generally darker colours and also adds strength to the feather. But such pigments are apt to fade with time – and especially with exposure to sunlight. Black parts fade to brown; brown fades to cream. Peter Hayman showed that a golden eagle grows a brand-new upperparts feather, for instance, that looks glossy black-brown, but within a single sunny summer season it can fade to pale buff. You can look at photographs of almost any bird and pick out old and new feathers – by their worn appearance at the edges and especially by their minute – or striking – differences in colour. Eagles often have two or three generations of feathers, and this can give them a blotchy look, a mixture of black, one or two browns and cream.

Pale grey birds are frequently subject to easily discernible change. Just look at something close to hand – a woodpigeon for instance has smooth blue-grey feathering that changes to dull brownish-grey with age, and it is easy to see newly grown feathers contrasting with older ones on the back and wings. Look at cuckoo photos and see if you can work out the stage of moult (tricky!) – are there bluer covert feathers amongst older, browner ones?

Robins famously fade from quite rich orange-red (but never Christmas-card robin crimson) on the breast to dull, pale, worn orange in the summer. Then by late summer, after a time when they are inconspicuous, keeping their heads down during their moult, they look as bright as ever once more. So, yes, the red of a robin is certainly a fixed and constant colour on new, fresh feathers, but it can and does vary – within limits – with the degree of wear and tear and fading it is subjected to.

So, colour – and pattern – provide birdwatchers, bird photographers and bird artists with endless interest, whether from a simple identification point of view (without them we would be in a real mess) or as a gateway into much more complicated matters. Ageing, the effects of moult, and especially all those magical and innumerable effects of local colour, lighting effects, reflections, highlight and shade and so on, all affect colour and are elucidated by it. Yes, a good deal of this is obvious enough, but like so much that we take for granted,

Can I come in? – a robin at the window on a wet day.

it offers a good deal of fun and fascination once you begin to think about the hows and especially the whys. The world of colour around us – and not just in birds – really is endlessly inspiring.

EARLY DAYS

One obvious truth about birdwatching is that you get very much better with experience, but there are two simple points to make. One, you *might* not get better anyway. Two, if *you* don't mind, who does? You might go for decades calling a kestrel a sparrowhawk, but if you aren't bothered, that's okay. Not everyone wants to be an expert. On the other hand, even if you have no interest in rarities, there is no reason why good birds should not pop up in front of you, before you even have a clue what they are. I saw two new birds on the same day when I was at school, one a redwing – an everyday autumn and winter bird that I'm now surprised I had not seen before – the other a great grey shrike, almost a semi-rarity. Because of the places I visited, I had already seen red-throated diver and black-necked grebe out on the water, and snow buntings on the shore of the local reservoir, while I was still struggling to be sure about redpolls, twites and reed buntings (largely, I suspect, because I was trying too hard to see a redpoll). I saw – *found!* – a November hoopoe in Staffordshire – completely wrong month, wrong place – when I was at school. Good job other people saw it too; it is still outrageous. If you go to the right kind of place, there *will* be birds. You just need to learn how to tell what they are.

My store of notes kept since I was barely 12 years old reveals a great deal about the way I developed and improved as an ordinary everyday birdwatcher, revealing obvious mistakes and moments of enlightenment, as well as reminding me where I went and what I did (but often, now, sadly incomplete – how did I get there? who was I with?). I began very much as a loner, with no-one else at school at all interested (so I didn't talk about it and really only had any interaction with others regarding birds by letter). Not only was it a solitary occupation, but it was not shared by anyone I came across more than a handful of times a year.

This random selection of extracts from my notes reminds me of good days and ordinary days and will, I hope, give you an idea of what birdwatching is about if you are just starting, what you might do for yourself, and especially how things have changed. Really, things were very different in the early years after I started: especially, the material that was available to help. My *Observer's Book* didn't do Caspian gulls and yellow-browed warblers. We didn't even have a phone in the house, let alone anything like a mobile. The internet would have been science fiction.

Even on into computerised days, an asterisk (*) always appears in my notes beside anything deserving a highlight – a good view, a good bird, a high count or whatever – and (New Bird) beside anything I had never seen before (except on many foreign trips, where these would be a bit overwhelming – go to Australia or somewhere and nearly everything is new, after all). When I was 15, 'new bird' appeared beside such things as pintail and black-throated diver (already making some good finds for where I was), even little owl: a random selection. These reflected where I lived as much as anything: there were three brent geese on the local reservoir after a storm (never to be repeated in my experience, so a good record) and these got the 'new bird' treatment. Had I lived in Essex, say, I would have grown up with brent geese. I did visit Essex a lot and had already, for example, been watching nightingales and corn buntings, arguably more difficult than pintails and brent geese. I recall one English lesson at school, well before O-levels, in which we must have read *Ode to a Nightingale* by John Keats or something, and the teacher asked if any of us had actually heard one, then, as I put my hand up, quizzed me as to whether or not it was anything special.

One memorable August day in Essex, when I had just turned 15, gave me the chance to see some more waders, on the estuary around Goldhanger. This was a wild and evocative area, some of the best of the Essex coast, the kind of place that featured in an RSPB film I was to introduce at shows many years later, *Wilderness is not a Place*, written by J. A. Baker of 'peregrine' fame. His book *The Peregrine* (1967) is widely rated amongst the finest of any wildlife writing ever done. His book extolled the values of the Essex marshes, where 'wilderness' was a creation of all our senses, the light, the sounds of wind and waves and birds, in which the coastal wildlife played such as pivotal part. I was not used to seeing as many as 250 curlews, but the 'new' birds (i.e. lifers – do people still say that? I never did, anyway) included both grey and golden plover, green sandpiper (flushed from one of the narrow borrow pits inside the sea wall, still in my mind's eye as a small bird leaping up to reveal its black-and-white effect in flight) and turnstone. Later there was a long-tailed duck on Abberton Reservoir (in 'eclipse' with few flight feathers, I noted), but this was not marked as new, although it would have been the 'bird of the day'.

I was fortunate to be able to visit the southeastern corner of Essex, where islands and hidden creeks and sea walls, farms and barns and combine harvesters and secret Atomic Weapons Research centres added a potent mix to family history and beloved relatives with memorable musical voices, and created a wonderfully different environment in almost every way from my own Midland home. It all has a kind of romantic air, now, with a certain gloss or maybe all-encompassing rose-coloured cloud around it, brought back to life by those old Box Brownie black-and-white family photos kept in a biscuit tin (and now lovingly digitised). How great it all was... wasn't it? On the islands, the land was all 'claimed' from the sea and so was, with the exception of the surrounding protective sea walls, entirely flat for mile after mile.

As I was to be reminded later on several foreign trips, 'flat' can – even for someone like me who loves hills – have an unexpected and powerful effect. There is a special sense of space and openness about flat land, with, as is so often said, such a vast dome of sky. Certainly, the area remains memorable and exciting: the different buildings, different habitats, different local accents, everything about it, together with sunshine and the temporary freedom of school holidays, creating a feeling of elation, eager anticipation and 'escape'. Not that there was anything to escape from, except perhaps from school. Nostalgia and rose-tinted spectacles add something, but little extra is needed. I am aware of having seen one bird when I was very small that, until relatively recently, I had not seen again in Britain, a night heron, which lingered for a while on an overgrown farm pond where my uncle worked. It is in the record books, but I have no memory of it.

Another new bird a few days after the Goldhanger waders, near Colchester, was a collared dove. Yes! This reminds us how recently (although it now seems a long time ago) the collared dove appeared in England. It is so familiar now, its three-note song and high, circling display flight everyday background sound and sight in most areas of Britain, but back in the mid-1960s it was still a rarity. (Like aeroplane trails in the sky, it is one to catch out television period-drama producers: it should not be in the background of a Jane Austen story, or even a postwar drama.) I remember looking for them around docks and distilleries, where grain was apt to be spilt, in my early Scottish holidays. Outside such places with guaranteed food, there were none.

Many other species have changed in status quite dramatically. Willow tits buzzed in the hedgerows as well as dense, damp, marshy thickets in Staffordshire, always good to see but easy enough to find; now willow tit has gone from much of Britain and is rare where it remains. Marsh tit is perhaps following it the same way although on the New Forest fringes its musical note still sometimes pierces the gloom of the understorey. There were turtle doves in any late-summer field after the wheat had been combined, even scores of them on overhead wires. I recall a count of 108. Even many years later when I used to go occasionally to Norfolk, migrating groups of turtle doves flying along the coast in spring were frequent enough. Again, turtle doves are now *extremely* localised and rare in Britain, reduced to fewer than 3,000 pairs. It is scarcely believable, yet some Mediterranean countries still allow them to be shot. Similarly, tree sparrows have gone from ordinary to very scarce, or entirely, sadly, absent, mostly. There has been a good deal of expert research into these declines, with a variety of conclusions either firmly or more tentatively drawn, yet sometimes I look at patches of habitat that still *look* ideal for, say, a willow tit or a nightingale (which likes its thickets low to the ground) and think they *are* still ideal, but the birds are gone. It is not easy.

A few large ploughed fields locally proved good for lapwings and I soon began to find golden plovers, too. Almost as soon as I started looking at them, I noticed that golden plovers had more or less conspicuous white wing-bars, contradicting my books, but was just beaten to it trying to put this into print. The same fields later began to attract breeding corn buntings, which were on a bit of a roll, back then, spreading and increasing through the midlands. The lapwing flocks have gone and in many other places are still going: in the past few months, as I write this, I haven't seen one at all in fields on the Dorset/Hampshire border, where

*Where are they now? The **tree sparrow** has withdrawn northwards and eastwards in recent decades and is hard to find, now, in much of the UK. It has fluctuated before, and this seems to be a 'natural' change rather than something related to farming or some other human influence. The **willow tit** has also more recently disappeared from many areas. I used to find it commonly in thick hedgerows in Staffordshire in the 1970s, for example, as well as damp willow thickets, and it is still there but in much reduced numbers. Researchers say that willow tits can no longer survive in woods where the undergrowth has been reduced by overgrazing by increased numbers of deer. Maybe so, but near where I live today, on the Dorset/Hampshire/Wiltshire borders, the woodland and roadside hedges are still often very dense with plenty of coppiced areas (which also look good for nightingales), but the birds are absent all the same.*

In my student days we would go to Tregaron in mid-Wales to see a red kite or two: seeing one of the 30 or 40 that remained was exciting. Now, following reintroductions, they are easy to see every day where I live in Dorset: still wonderful, if no longer giving quite the same burst of adrenaline. But I still especially like the buzzards. Never so rare, but still we had to go to Wales to find one in the 1970s, before a welcome resurgence in the Midlands and east. Oddly, they seem to have declined on the Dorset border recently, perhaps because of a drastic reduction in rabbits. I was much more puzzled than I should have been, or expected, when I came across a black kite in the New Forest a while ago. I had, after all, watched scores of them abroad, and they were never any trouble. Fortunately, this one came closer eventually and showed itself for what it was.

only a few years ago there might have been hundreds if not thousands, with the odd ruff or other hanger-on.

It was early in 1966, though, that another 'new bird' came my way (after such good ones as snow bunting and waxwing) – my first sparrowhawk. Birds of prey in the early 1960s were in a parlous state – buzzards were restricted, red kites almost extinct, peregrines very local and rare, even sparrowhawks scarce and hard to find. The reason was pesticide use on farms: DDT and related substances. In America, Rachel Carson had written *Silent Spring*, and as we have already seen suffered a great deal of abuse too from all sides, being a woman for a start, poking the finger at giant multinational companies and their 'magic' pesticides. The book pointed out the dangers of these chemicals and became a conservation classic. Few people know about it now. Some things that, to some of us, seem so obvious are unknown to the majority of people, who have no real idea about this difficult and dangerous period, when hundreds of thousands of small birds died after feeding on contaminated seeds, and then predators that ate them also either died or failed to breed.

Eggshell thinning was a strange side effect of organochlorine poisoning from DDT, aldrin and so on (you could buy little packs of aldrin and dieldrin: I used them myself when helping out in the garden). This was studied in Britain and one of the scientists involved later became the director general of the RSPB, and was therefore, several layers removed, ultimately my boss – Ian Prestt. But back to 1966 and the sparrowhawk. It was like seeing a rare vagrant, quite unexpected: nobody saw them then, in south Staffordshire. In September 1967 I described in my notes my third ever, but I clearly had some difficulty with it as it dashed by – not quite the shape and flight I had expected. And then December of that year brought a 'possible', what would have been my fourth, but better views were needed. I was especially cautious because I was with two much more experienced birdwatchers – neither of whom had ever seen one before.

Since then, things have improved for birds of prey and they have increased hugely: but have lessons been learned well enough? There are still huge impacts out there on other species, from a variety of farming operations. Modern pesticides seem likely to have just as big an impact as the old DDT did, yet they repeatedly get licensed for use, despite massive opposition. A friend of mine worked for the government environment agencies and warned of the dangers, but was told to keep his mouth shut, he was sailing too close to the wind. A government minister recently (2023) explained on TV that bees did not get close enough to the pesticides to be harmed, as farmers were encouraged to keep wild flowers away from their crops. What does *that* mean? Watch this space.

At the same time, I noted 'many skylarks singing – 150?' This was not the only occasion that I estimated 100 or 150 singing skylarks on my walk of a few miles to the local reservoir and back. Could I have been wildly overestimating? It seems hard to believe now, but there were, obviously, several scores of singing skylarks. on a mixture of agricultural and pasture land and rough heath. On another day I noted 20 singing skylarks along one quite short piece of heathy shoreline beside the reservoir, but sadly have no properly documented counts. My birdwatching needed to be more purposeful! Recently I have walked for many miles through scenically attractive farmland in different parts of Dorset, Hampshire and Wiltshire, and usually manage one, or two, at best perhaps five, singing larks over a several-mile walk. Of course, there are places with good numbers, still, but the old days of skies full of larks seem to be long gone.

It is interesting to see in my notes in the late 1970s that winter flocks in this bit of Staffordshire had perhaps 400 skylarks, quite remarkable now, as well as maybe 50 yellowhammers. Older bird books talk of 'stackyards' where there would be flocks of finches, buntings and sparrows, as well as pigeons (but not collared doves!) as well as rats and mice, attended by sparrowhawks, kestrels and barn owls. These went long ago but the flocks of birds remained in autumn/winter fields. How often do bird books today refer to 'stubble fields'? Can you find a decent stubble field now? Stubble was the remnant short stems (complete with spilled and lost grains) left after a cornfield – wheat, oats or barley – had been harvested. In 1966, in a quite ordinary little field, just left to go a bit overgrown, I was watching 200–250 greenfinches, hundreds of house sparrows and an uncertain (not noted) number of linnets, plus some reed buntings and yellowhammers. It was a 'normal' mixed flock. Not far away, a couple of weeks later, I watched 50 chaffinches and 100 tree sparrows: I underlined tree (and added 'few house sparrows') as that was pretty unusual even then, although I saw tree sparrows quite often.

The same little weedy field in 1967 brought my first corn buntings locally: the West Midland Bird Club's 'field meetings officer', Arthur Jacobs, pointed them out to me and said they were the first he had ever seen in the Club's area (Warwickshire, Worcestershire and Staffordshire, including what has since become the West Midlands conurbation). They were with 30 reed buntings, 300 greenfinches, a few redpolls, 100 house sparrows and some yellowhammers. The field also attracted my first collared doves in the area. In later years, breeding corn buntings appeared in nearby fields and built up to 22 singing males. There are none now. Things change.

The stubble field was accessible during a period that reflects our recent trials with Covid, when much of the countryside was out of bounds because of foot and mouth disease. In December 1967 it was cold, windy and often snowing. One Saturday, from the house I saw around 200–300 greenfinches, 20 chaffinches, 100 house sparrows and several bramblings fly over, fleeing the hard weather. The next day on a walk I noticed similar numbers, flying west, as well as a mixture of fieldfares, skylarks, starlings, woodpigeons and lapwings. When I reached the little stubble field a massive mixed flock whirled round in a 'roar' of sound, both calls and wing noise, before feeding in bits of uncut wheat and

A few hundred house sparrows whirr across a wheat field. There are still such flocks in some places, but they are harder to find now than they used to be. Around the Hampshire/Dorset border I find small numbers in the villages inland but a better situation near the coast, where gardens butt up against patches of rough grass and overgrown earthy cliffs.

The giant, sculptural elms are also a thing of the past, following the ravages of Dutch elm disease, a virulent form of which reached the UK in 1967. Millions were killed. As a child, I was told not to go under them, as they had the habit of dropping branches unexpectedly: perhaps this was from some hard-earned, long-ago experience? Anyway, it is as well to remind people, and to tell those too young to have known them first-hand, about such things – the landscape of the English elm was often so very different from the current one, trees punching up into the sky in shapes like miniature editions of soaring cumulus clouds.

long stubble. There were 60 skylarks, 40 reed buntings, 40 chaffinches, a few bramblings, 400 greenfinches, 400 house sparrows, a few – not worth counting, then – tree sparrows, 10 yellowhammers, 200 starlings and some collared doves.

Another flock struck me as particularly fine, as it happened to be mostly male chaffinches coming into breeding plumage (as their bills turned blue and the dull edges to feathers wore away to reveal bright colours beneath). The scientific name, *coelebs*, means 'unwed' or 'bachelor', and many chaffinch flocks are indeed composed of males or females and are scarcely mixed (although flocks feeding on beechmast in woodland, for example, often seem thoroughly and evenly mixed).

With these few hundred male chaffinches were quite a number of bramblings; in another flock, there were 100 bramblings and fewer chaffinches. My best brambling flock locally was 440+. They reminded me of a large flock of snow buntings, with so much white and complex contrasty patterns, but with much of the white replaced by orange. In an area of southwest Staffordshire that was not known to our family until I visited to see some birds, there were regular records of ravens and buzzards, otherwise virtually absent from the county. But it was when I heard of flocks of bramblings over there, in the Enville/Kinver area, that we went to have a look and did, indeed, find a field with *hundreds* of them. Alan Moffett, a local bird photographer who had several features in RSPB magazines, later photographed goshawks there, as well as his brilliant cuckoos and hawfinches. Thinking of him reminds me of another change: around 20 years ago he wished to donate his collection of 35 mm transparencies to the RSPB photo library, but by then they were not wanted. Everything had gone digital. It was an unfortunate if understandable rebuff to a generous offer.

These singing skylarks and mixed finch, bunting and sparrow flocks were in very ordinary places, not managed or protected for wildlife in any way. Now, finch flocks are few and far between in general: younger birdwatchers perhaps don't realise what ordinary farmland birdlife was like not so long ago. Similar fields in the Trent valley, in summer, would also have breeding lapwings, redshanks, snipe and memorably curlews, too. These have all gone. Curlews, in particular, have been declining everywhere for years. I now live near the New Forest, originally designated an important bird area, even a Ramsar site, for its breeding waders (snipe, lapwings, curlews, redshanks). The Forest is brilliant, still, but the waders are nearly all gone: dog walkers include a minority (small I hope) who are determined to ignore pleas to control their animals: 'It's a free country, I can do what I like, I can and will let my dog run where it wants. By the way, I pick up after my dog, even though I then throw the plastic bag into the grass or hang it from a tree, or put it neatly beside a gatepost for someone else to dispose of.'

It was many decades ago that the corncrake was lost, so that means little to me; but it is sad that the loss of lapwings and curlews and redshanks from such places will mean nothing, now, to modern birdwatchers. That's the 'shifting baseline syndrome', thinking that the diminished birdlife of today's farms is 'the norm' rather than the riches of 30, 50 or 100 years ago. With so many problems in the world, the state of the curlew population in lowland England might seem of no great consequence – but it is a sad situation and getting worse.

Many of your earliest birdwatching experiences will stay with you. I read the *Ladybird Book of British Birds* as well as the *Observer's Book*. Some books still use rather quaint old 'bird book' language that echoes these, which I try to avoid (a bird's habitual haunts, breeding grounds and so on). But there are impressions made by those books that will never smooth out and disappear. In particular, I still remember the illustrations and especially just the lovely names of some of the birds that I wanted to see – and I still think yellowhammer and bullfinch are just fabulous birds, and brilliant names both. 'Yellowhammer' just sums everything up, for me, about early birds, common birds, beautiful birds – nothing unusual exactly, but just a bird that is a joy to see and to hear and to write down. Nowadays, a relief to know it is still there. We used to find their irregularly lined eggs: we called yellowhammers scribblers. Or maybe older boys did and I just followed suit. I don't suppose I knew what they were at all, as my birds'-nesting days stopped before I was 10.

We must clearly be thankful for the efficiencies that feed our burgeoning population, but vast acreages of modern farmland are now poor for wildlife of all kinds. The average stubble field is very short-lived and often *absolutely birdless*, not even so much as a woodpigeon or a crow. The only interesting corners, otherwise, are left to help birds that will later be shot. The RSPB has long worked a Cambridgeshire farm to experiment with methods that still yield

Barn owls need rough grassland or marshy fringes, where they have a chance of catching a few rodents. Fields cut for silage or heavily grazed just don't do it for them.

good harvest but give wildlife a chance. For a time, it seemed that recommendations to farmers would be moderately well received and taken up in fair numbers, but looking around now, I see no evidence that anything much is being done except for yet more pheasants! Down come the lines of Range Rovers, the guns are handed out, bang bang bang for a few minutes, and they move on.

Another area of less squeaky-clean fields where I sometimes walk generally has rather few birds, but one year (2013) was a draw for birds and birdwatchers alike. A bit of rough downland clearly had a burgeoning vole population. Where now there might be a kestrel, there were several short-eared owls, one or two barn owls, kestrels, buzzards, a hen harrier... not far away a couple of red kites and, no doubt coincidentally, a goshawk. And in the bush-tops, a wonderful great grey shrike (that is, an ordinary great grey shrike, a species that happens to be wonderful).

Here's how I recorded one visit, by now in a Word document on my computer:

> Persistent light rain and cold seemed to have spoiled the visit and we were about to leave after two glimpses of short-eared owls, but 'gave it another five minutes'. After that for an hour or more we had consistently brilliant views of short-eared owls, often very close, sometimes on the ground briefly (when I could see the 'ears', which is unusual), ranging widely both sides of the road. One was especially pale but there seemed to be three pale, one moderately dark and one particularly dark, a rich tan-tawny bird with a lot of dark brown on the upperwing, quite dark even around the face and underneath, a less striking bird at a distance but very handsome close up. The medium-dark bird had three dark tail bars I think, paler birds four (possibly five – surprisingly hard to count) but this dark one seemed to have more narrow dark bars on a finely banded tail. It could almost look like a long-eared except for the yellow eyes and the white trailing edge to the wing (although that was relatively weak and narrow). The palest bird was wonderfully striking, especially head-on, with broad white leading edges to the wings and white marks at the base going onto the 'shoulder' spots. I heard one call, a short, downward *aow*.
>
> Fieldfares in a treetop group set up a prolonged fast, jumbling, chattery chorus of subsong mixed with calls, for 15–20 minutes at least – I don't recall hearing this before.
>
> I looked at a group of starlings flying over, and a peregrine came into the binocular field of view – I just missed by a fraction of a second the actual

> contact, but it took a starling very neatly and easily in a downward sweep, and flew away with it, looking very large and long-winged – later another, this time more likely a male, flew over going east – a good, long view.

The thing isn't so much *why* all those predatory birds were there that winter. Voles, presumably. But I wonder *how* they came to find the place. How come these owls, raptors and a shrike all homed in on a rather small patch of grass? Are we to think that they are all passing over, anyway, year after year, but usually don't bother to stop? Just an invisible movement of exciting birds that we never see? How long before they realise there are voles to be had, if they are just flying by? It takes just a few moments, after all. It sounds so unlikely – but how else? Perhaps they have their smart phones and social media.

Incidentally, there were eight Mediterranean gulls, too, circling with black-headed gulls and drawing attention to themselves by their calls. They may have been feeding at a nearby pig farm. Good gulls get everywhere...

Quite ordinary birds can produce extraordinary moments of course. Here's something I have always remembered, partly because (still in my early teens) I naively wrote to Alan Richards, secretary of the West Midland Bird Club and the man who did the monthly newsletters, and told him about it along with a few reports (as if he hadn't got enough to do). Here's my notebook entry:

> Four kestrels flew close together, one adult male. I sat on top of a large pit mound [a spoil heap of mining waste] and three young kestrels used the mound like a fulmar uses a sea cliff – hanging in the strong wind that blew against the slope. I had two, sometimes all three, within 50 yards and at times as close as 25 yards for more than 10 minutes, in bright sunlight. Being newly fledged, they were in immaculate plumage: similar to an adult female, one or two of them with a greyer tail, but the feathers were much smarter than the rather ragged adult. It was a beautiful sight to have a kestrel hovering 30 yards away seen through a telescope at 60x magnification. Several times one would hang in one spot for a long time, absolutely motionless except for an occasional flexing of the primaries and twitch of the tail. Every detail was visible, down to the scales on the feet. Most views were from behind and to one side, looking along the tail and rounded body, shaped like a long spindle. The feathers lifted and rippled in the wind, looking soft and owl-like.

Given the number involved and the 'immaculate' plumage, I daresay I was right, but telling a juvenile kestrel from an adult female is not always easy and I don't think I knew or used the diagnostic features, such as they are. And there was too much of a 'look how good I am, being interested in common birds, not just scarce ones' about it all.

How about a seawatch?

Might you have the chance to get to a coast? If so, it is worth trying a bit of 'seawatching'. It can be what it says on the tin – watching the sea, and not a lot moving over it. Or it can be dramatic and exciting, with all kinds of seabirds on the move and visible from the shore. Usually, it is somewhere in between, but it depends where you are and especially what the weather is like. Find a headland, or a bay between headlands into which seabirds can be pushed by strong winds. They then have to squeeze by to get out of the bay, passing one of the outer bulging shorelines. Birds that prefer to be out to sea will then be much closer to the shore than they are comfortable with, but give far better views.

Amongst the classic seawatching places are Liverpool Bay, famous for its Leach's petrels, where the birds come in and must sweep out to the west, passing close by the end of the Wirral and other prominent shores, and St Ives in Cornwall, where they come in close, head south and must turn back west to get around the awkwardly placed headland.

Take a notebook! You can't remember all the numbers as things chop and change and new birds are always moving by. Keep notes. And be ready to describe something, properly, written down, so you can be sure, if you can't identify it right away and need to check it properly later on.

Several species rely on wind, or air currents over the wavetops, to carry them along with little expenditure of energy. Being forced against a shore gives them little room for manoeuvre. Many headlands will prove worth visiting for passing shearwaters – 99% Manx shearwaters but with sooty and others always a possibility – fulmars, gannets and skuas. Most of the skuas will be great or arctic skuas, but again there's a chance of pomarine, even a long-tailed skua, to liven things up.

Perhaps there will be a Sabine's gull amongst the kittiwakes? Sabine's come from the Greenland area but there is a tiny population in Svalbard, which may well head off east around Asia rather than come our way. Or there could be Leach's petrels brought in close enough to show their angular wings, bounding flight, pale upperwing bands and white rumps. Later on, say early November,

Always high on the agenda for a bit of seawatching will be arctic skua: one of the most perfectly proportioned birds of all. Kittiwakes are often unfortunate targets, being forced to disgorge food that the skua will then steal for itself.

northerly winds in the North Sea could bring little auks passing places such as Flamborough Head, or squeezed down to the Norfolk coast.

The point is, you never know: there's always a *chance* of a good day. But be sure to find a comfortable, sheltered spot where you can sit without getting wet; take that notebook and pencil (pens don't work so well on damp paper), maybe a fisherman's umbrella, and be ready to focus calmly, objectively, but quickly on whatever comes by. If you are with others, try to establish some landmarks – maybe a buoy – or a 'clock face' idea (12 o'clock is straight out) so you can shout out what you are looking at. *Be quick, for it will not be there for long.* Seabirds have a habit of flying by and never coming back.

But you never know. One September day, I just sat on a sandy beach and had a wonderful time:

Avon Beach, Mudeford, Christchurch

**Grey phalarope	1 juvenile/1st winter transition
Turnstones	15–20
**Sabine's gulls	2 1st winters
*Little gull	1 juvenile/1st winter – still in between
Sandwich terns	c. 6
*White wagtails	2 juveniles/1st winter

Three superb hours: an hour or so in dull but clear light, then bright sunshine – wonderful birds along the beach with groynes, the gulls and phalarope usually within a few yards of the beach or over the rolling and breaking waves.

Grey phalarope – the best I've seen in some ways, pretty close at times, but feeding actively on big swell and breaking waves instead of on a flat inland pool. How they should be, somehow. Superb views. Still much juvenile plumage above, with big grey scapulars coming through, and a buff wash across the foreneck and upper chest. The down-tilted bill, following the curve of the throat and chin, and perhaps the small paler base (?) give an illusion of a faintly downcurved bill. In brief flights with dangled legs, it seemed that the lower legs and feet were sandy-ochre. Sometimes seen swimming within a few feet of a Sabine's gull. It was around all afternoon.

Little gull – marvellous close views – still a lot of dark colour on the hindneck, chest sides and rump area. Brilliant views of the plumage detail. Wingtips slightly rounded; noticeably smaller than Sabine's, with quicker wing action in steady flight in calm air.

Sabine's gulls – magnificent. I don't know when I last saw one – years ago! Many very close views in flight and on water; one finally resting on the sea or bathing, but mostly feeding actively, either in flight or swimming and picking from the surface.

Very dark on the water; shape much like black-headed, with long, fine, upward taper to wingtips, but head often held relatively higher, and proportionately a little smaller – head looked very round but closer views showed long forehead and 'snout' effect, before short but finely pointed bill (clearly thicker than little gull's, but still fine and sharp). Much smaller than black-headed, but in flight seemingly large-winged – basal half broad, but perhaps exaggerated by the pattern. The wing area presumably helps give extraordinary aerial agility with deft, sudden, rapid twists and falls;

perhaps also aided by the extra sharply angled carpal joint. Steady flight in calm air direct, with soft, padding action. Head-on, very pale-headed. On water, dark with white face, foreneck and chest creating small patch at front. Superb views of all plumage patterns. Both much the same as expected, but one had about three rounded, blacker 'fingers' within the grey chest patch on the left side; the other was cleaner and paler. The dark bird may have had lead-grey legs; the other seemed (but in bright sun) to have paler, pinker legs.

Quite often heard calling, very quietly at close range, with an insignificant, sharp, metallic *peep* or *keek*. Beautiful, striking birds.

This was a great afternoon: hardly knowing what to look at first. Fantastic.

As a rule, though, seawatching is a different matter. North and west Scotland on my teenage holidays used to be good for local species in high numbers – gannets, auks, kittiwakes, fulmars – but from time to time there were also sooty shearwaters, never easy to see. From the north coast of Norfolk in a November gale I have enjoyed little auks, made that little bit easier in this case by the sheltered touristy beach-side benches. From eastern England I didn't so much enjoy seeing other people watching a long-tailed skua, which had gone by the time I had reached the cliff top. But the chance of an arctic or great skua always speeds the adrenaline flow that little bit.

A fair wind – mid-August, Cornwall:

Trevose Head, Cornwall

**Manx shearwaters	15,000+
*Fulmars	c. 30
**Gannets	750+
Shags	c. 10
Kittiwakes	3–4
Peregrine	1 juv
Wheatears	3

I would have been happy with a shearwater or two, and a few gannets… I looked out and immediately saw the sea swarming with them. Local seawatchers said they estimated Manx shearwaters passing at 12,000 an hour, but my sample counts were 200 per minute two or three times, one of 300 per minute and one of 500+ per minute, so I think this was substantially underestimating the true numbers – I could have seen many more than 15,000, but watched a little bit haphazardly as we walked around and

> enjoyed the place and the views as well as the birds. Gannets were just fantastic, too – recalling (as ever) W. H. Hudson's writing about watching the passing gannets by the hour (from Cornwall).
>
> Shearwaters gradually reduced later, but then a group of 200–300+ settled on the sea relatively close in – and they were chased repeatedly by a peregrine, diving and swooping at them and making them, in turn, dive into the waves with big white splashes.
>
> I have never seen Manx shearwaters in such numbers anywhere before! Amazing.

One October, with my old mate Peter Garvey, I had a few days in Pembrokeshire. I think we hoped it would be a bit of an adventure, with a few rare American migrants to discover... it didn't happen but we saw some birds, nevertheless.

We were seawatching off St David's Head. Apart from 25 gannets, there were really no seabirds at all. A couple of choughs made up for it a bit. On a nearby beach was a grey seal with a tiny, pure creamy-white pup. But there was a bit of an oddity, too:

> While seawatching off the head, I was suddenly confronted by a bird flying low over the sea that brought a touch of the exotic (not to say ridiculous) to the occasion. It was a superbly beautiful flamingo, probably a Caribbean (American) flamingo. It was a bright, rich pink, especially on the forewing and body but nevertheless quite pink on the head and neck – nowhere white – with jet-black flight feathers, and the whole of the body looked reddish-pink. The shape, with exceedingly long neck and legs, was quite incredible and it looked a really beautiful sight.

Most flamingos seen in the UK are escaped Chilean flamingos, which are much duller and have grey legs with pink 'knees', the clear distinction from European greater flamingos. Greaters are less vividly pink overall than the bird I described (although with stunning crimson and black wings). So this did indeed look like a Caribbean flamingo, and must presumably have been an escape, despite the strange location. Not a wild one, surely, was it? Was it?

Wales gave me a couple of other 'escapes', including a saker falcon on the Berwyns and, most interesting perhaps, a splendid male lazuli bunting, in song, early in summer. I've often wondered about that one, but it is a western bird in North America, not really likely to make it to Europe. Is it?

A 'brown iris' view

With gulls, I have sometimes noted a 'through the nostrils' view, when you can see a bird so well that you get daylight showing through the nostril from the other side – pretty close. Snipe watched with a telescope from a hide at Blashford Lakes, feeding on the shore beneath the windows, have often given the kind of view that allows an easy separation of the brown iris from the black pupil in what would normally look like a 'black' eye.

One December day back in the early 1980s brought such a view of a different bird, this time at Cambridgeshire's Little Paxton pits.

> The ground is still ice-bound and largely snow-covered. Today was bright and clear but there was little warmth in the sun and no sign of a thaw. Apart from blue tits and hundreds of woodpigeons, the area I visited was relatively birdless.
>
> An unusual spot to park the car was a patch of leaf-litter kept clear at the edge of some trees: and there right in front of me was a woodcock! It ran away a few yards but stopped and looked back. After a minute or so it returned to the leaves, crouching down and pushing its way through some brambles back into the open in bright sun. Phenomenal views, from about 12 yards! I've never seen a woodcock like this before. It looked large, bulky, broad-backed and low-slung, on short but thick, pinkish legs. It seemed generally rather dull grey-brown and with the pale areas on the upperside greyish rather than clear buff, the head pattern less contrasted than I would have imagined and the barring below rather pale (with the upper breast more solidly brown). Apart from the complex, intricate patterning and

> the long bill, the chief features that stood out were the eyes – set very far back and high on the head, from some angles almost as if sticking out on 'corners' and huge. They looked blackish but good light revealed very dark, rich brown irides.

Years ago, Keith Vinicombe made fun of me pronouncing this as 'eye-rides' instead of 'eye-rid-ees'. I'd never heard anyone say the silly word...

The weather can be good anyway

> Exactly what I hoped for has indeed happened. An overnight clearance of cloud has given a morning, after a hard frost, with no wind but blue sky and bright sunshine, which turns the snow from grey to brilliant white and blue. The snow obviously came down in just enough wind to drive it into every nook and cranny, but not enough to blow it away when frozen hard – so every tree, every bush, every hedge, even dense evergreens, have snow right through to the centre on every branch and twig. It is not just a superficial covering, nor the thin, delicate tracery of a frosty rime, but anything up to three inches of snow along every spray, so the trees look like huge, complex white corals or infinitely delicate white cauliflowers, or vast chrysanthemums!

My notebook was not usually quite so poetic as this extract from a December day in the 1980s (at that time still a handwritten account, predating my change to the computer). It continues:

> A superb day, bright and sunny still, full of good birds and interesting things to see. Several of the pits were completely iced over; two practically free; the largest 95% frozen with all the coots [1,300], gadwalls [185] and many others mixed and packed in the free space, making accurate counts difficult. [*340 mallards, 30 teal, 60 wigeon, 215 pochards, 260 tufted ducks, 14 goldeneyes, 11 goosanders, a male smew, 63 mute swans.*] In a small area of open water on the far side was a good group of tufted with odd gadwalls and wigeon. Later the tufted flew off but returned in two groups along the length of the pit – almost leading the second bunch was the drake smew. Later I saw it briefly on the water before it flew back past me. The goosanders were also flighty and it was difficult not to disturb them. I saw a kingfisher really well, one of those rare (for me) occasions when all the 'blue' areas looked pure green, even the stripe under the face. It moved

> several times then flew to the river. Almost immediately afterwards I saw the sad sight of a dead kingfisher floating face down in the river, but two more very much alive flew upstream, calling.
>
> In quite deep snow, as I stood still counting the ducks, I noticed a female blackbird just 8 feet away in the snow. It flopped down, looking at me and called repeatedly. I bent down and moved closer but it fluttered then flew into a nearby bush. A little later, it called again and flew around me, so close that I thought it was going to settle on my shoulder. On a flat field with several inches of snow, there were odd dark patches that proved to be root crops and clods of earth. Each of these – both roots and lumps of soil – had been pecked out by rooks and skylarks, and scattered skylarks could still be seen [75–100], some half-hidden as they crouched in hollows in the snow, finding a meagre living somehow. Tracks of foxes weaved across the field and they 'joined the dots', going across the bare patches to each clump where a bird had been. Fox tracks were everywhere, including some up a bank from the river to a spot where a few scattered feathers indicated a kill.
>
> The river seemed to have no current and its surface was flat, mirror-like, reflecting the snow-covered banks, snowy trees and white mute swans above – phenomenally beautiful.

I was very often moved by such days and such scenes, but rarely had the time to write quite so much in my bird notes. Perhaps, after all, it is best to live it to the full at the time and not to rely on notes or photographs.

Other days, though, gave different effects that were equally remarkable in their way. One December, there was an area of long, tussocky grass, coarse and yellowish and quite ordinary, with a mixture of old thistle heads and the like – nothing much. But against the light of a low, orange sun, it took on a remarkable appearance. Over the whole area, a few hundred yards wide perhaps, every stem, every thistle, every bit of grass was covered in a tangled mass of the finest cobweb, mostly oriented roughly north–south, so that against the light a fine mist appeared to cover the ground. Binoculars – so often so useful – resolved everything into innumerable fine, silky threads draped from the stems. Only a few days earlier there had been a hard frost with thick, freezing fog. The afternoon was briefly bright and sunny, but the frost persisted in the shade and the mist returned by evening. I crossed Cannock Chase, which was 'out of this world' – every bit of grass, bracken and heather, every finely branched birch tree, was white with frost, with an incredible fine, intricate network of grey-white against the blue sky. It was a joy to be out, despite a complete lack of birds. One day there was a similar effect but neither frost nor cobweb had any part to play:

it was simply uncountable tiny twigs of birch, each shining silver against the descending sun.

Even in dreary rain the landscape can be remarkable. Again one day there was fog on Cannock Chase, but steady rain began to wash it away. Within two minutes I was watching a great grey shrike, but the rain intensified and there was nothing else to see as I simply got wet in the downpour. It was in its way magnificent, with yellowing bracken, yellow, orange, rust and tawny birch, beech and oak, all with remaining leaves having reached a perfection of late-autumn colour: a world of rich, glowing, yet subtle and harmonious colour. And rain.

And of course the English Channel coast in Hampshire and Dorset has its moments. Here are five different days:

> A beautiful afternoon, with warm sunshine between spells of heavy cloud and some torrential downpours, including a little hail, and even one thunderclap – a truly rare event down here! Remarkably clear (especially the Isle of Wight in bright sunshine), one of those unusual days when the very tops of extremely distant big cumulus clouds could be seen right down along the skyline. Some great downpours when the sea seemed to change – although there was a very light breeze off the land, there were some big breakers rolling in at times, but during the heaviest rain the sea smoothed out but showed many very long parallel lines of big waves rolling in to the shore – not just a trick of the light, I'm sure. After rain, the tarmac paths were producing clouds of rolling steam. It was superb weather, not matched by the birds, as ever on this bit of coast.
>
> ...
>
> Despite recent near-daily reports of all four skuas, for instance, again a nearly birdless sea. Nevertheless, I could hardly complain as, with heavy showers, longer spells of rain, and dry times with some blue sky and sunshine, the whole vista of the bay from Purbeck to The Needles, as well as the sky inland, was just spectacularly magnificent with wonderful lighting and sky effects – very dramatic at times.
>
> ...
>
> A dull morning with a lot of cloud, very soft, mild, but very clear to the west and south, Hengistbury Head and the higher parts of the Isle of Wight rising clear from a smoky haze to the east; the sea very calm and silvery. A very ordinary, very beautiful day: sunny later, but I think the best part of the day was the cloudy morning.

...

A beautiful day. Began grey and dull, but became sunny on the coast with scattered low cloud; a good deal of cloud and mist largely enveloping the Isle of Wight, The Needles generally clear. Sea flat, like polished pewter; very high tide, the sea inside the spit silver and blue and reflecting creamy white clouds, between belts of low saltmarsh; the grey bulk of Hurst Castle and the white lighthouse soft, floating, in the distance. Almost like a cold spring day as we walked out to the castle and everything became sharp and bright – but then the grey cloud won the battle and rolled in from the sea and we walked back along the spit in our own little world of cold, wet, grey fog, surf, faint brent geese and the double blast of the Needles foghorn.

...

Sunshine before and after but mostly a grey, cold walk – but for a time a break in the cloud above and sun shining through thinner cloud nearby, giving a curious lighting effect – the narrow lines of foam on the water's edge were really vivid, clean white against pale grey-green, and even when I used a handkerchief I noticed it seemed to be brilliant, almost luminous, faintly purple-white.

I have always thought that skies make one of the finest free shows around, as these extracts from past years make clear. So it is not just increased awareness on my part, but it does genuinely seem to me as I finalise this book that the year 2023 has seen an exceptional succession of wonderful skies and cloud effects, day after day.

Hard weather, hard times

Given access to food, most birds can survive cold well enough. Many, though, are simply unable to feed if there is snow on the ground or hard frost encases everything in ice. Obviously, birds such as grebes and kingfishers cannot feed if lakes and rivers are frozen over; water rails can't poke and probe about if the water's edge is ice-hard (they can then become surprisingly predatory); snipe can't probe into mud if it freezes solid. But nor can lapwings, or larks, or redwings and fieldfares, feed in the fields. They have to move, quickly. You might see an obvious 'hard weather movement' as many birds go west and south to find milder conditions. Or, if you watch a lake or reservoir with a patch of open water (perhaps kept free by mute swans and milling, diving coots), you may find something unexpected turning up for a day or two.

It started cold and snowy in 1967. A couple of thousand starlings flew over the house, oddly line-abreast, not single file. I went to the reservoir but it turned out to be still and foggy. Large numbers of gulls were on the water in the daytime, unusually, presumably affected by the fog. It was otherwise disappointing, very cold following snow but with no obvious influx of birds. Clear, frosty and snowy weather changed to miserable cold, dank fog, so anything farther off than 20 yards became a grey silhouette and, beyond 50 yards, hardly visible at all. The ground was iced over, a thin, glassy film.

Next day continued very cold and there was a good deal of ice on the water. A few hundred woodpigeons flew northwest, then southeast, so didn't really have any obvious aim in mind. Next day there were redwings and fieldfares, very approachable, on the school playing field, as well as a couple of thousand starlings. By the weekend, things were really on the move: hundreds of finches (probably chaffinches mostly) and then two flocks of woodpigeons, each of 2,000 or more, moving steadily west, determined, plodding refugees, unusual here. Over the next few days, thousands more woodpigeons in smaller flocks, all going west towards Wales or Ireland or southwest England, wherever there would be a break from this snow, ice and freezing fog.

This was a small movement really, but obvious enough. In south Wales once, while I was busy attending lectures, writing essays, whatever I should have been working at, I couldn't help but notice birds going overhead all day long: many thousands of lapwings, golden plovers, woodpigeons, skylarks, thrushes and various finches. Probably 10,000 redwings. Wildfowl and waders would have gone unnoticed, perhaps at night. I remember one of my friends had been in touch with a Cardiff birdwatcher (was it Amy Heathcote?), who told him that her nephew had had a large movement in the night. We hoped she must have meant birds, of some sort.

At such times birds can turn up in odd places (just as they do after autumn gales) and species such as redwings appear unexpectedly in gardens, or in the little shrubberies around supermarket car parks, where they are tame all at once, or really too tired and unfit to fly off as they usually do. They give us good views, but for them it is a critical time.

Unseen, small birds such as wrens, goldcrests, long-tailed tits – and rare Dartford warblers on southern heaths and Cetti's warblers beside rivers and marshes – die in large numbers and their populations may 'crash' in severe weather, although, all things being equal, they are able to bounce back again quite quickly given a good breeding season or two. Even larger species such as grey herons are affected, with periodic serious declines.

For some birds we can help by putting out food; others just have to struggle through.

A different kind of movement caused by the weather affects seabirds. I co-wrote an article for *British Birds* assessing the effects of the great storm of October 1983. It was easy to gather records and summarise them but, at the end, a thought occurred to me. Obviously, most seabirds on the coast stayed where they were: flocks of herring and black-headed gulls, cormorants and the like. Scarcer ones, from farther out at sea, such as petrels, Sabine's gulls and phalaropes, seemed to be selectively driven inland. The remarkable thing was that, by the very next day, there they all were over lakes and reservoirs scattered across the UK. How did they manage to find these, often relatively small, patches of water, so quickly, amongst so much dry land? None would have had any experience of following rivers or any other inland features before, but it is hard to imagine a Leach's petrel, for example, flying high enough to be able to detect the gleam of water on a dark night from a very long way away.

Every one counts

In January 1979 I was unexpectedly asked by James Cadbury to accompany him on one of his regular Ouse Washes wildfowl counts. Similar counts are made by hundreds of people across the UK and coordinated into an invaluable and unique ongoing survey of ducks, geese and swans, the BTO/RSPB/JNCC Wetland Bird Survey (WeBS), which goes back as far as 1947. Dr James Cadbury, tall, soft-spoken, gentlemanly, was head of the RSPB research department at that time, and senior ecologist with particular emphasis on the society's nature reserves, widely respected and popular. I was honoured to be asked to help him.

The Ouse Washes stretch diagonally across East Anglia in a long, straight line northwest of Ely, whose magnificent cathedral dominates the distant skyline. They were created to take excess water from the river Great Ouse, with two dead-straight, parallel channels, the New Bedford and Old Bedford rivers (it was the Duke of Bedford who was responsible), cut across a huge bend in the old river. The drainage works date back to 1630, and in 1649 a Dutch engineer was engaged to finish the project. The resultant water meadows and marshes and various shallow channels have been divided up into nature reserves managed by the Wildfowl & Wetlands Trust, the RSPB and the Wildlife Trusts. Few areas in England are so important for birds. Traditionally, the place was known for its ducks, Bewick's swans and breeding black-tailed godwits, but there is so much more.

Ouse Washes – Welney

CORMORANTS 2 *MALLARDS 3-400 *WIGEON < 500 TEAL Few
SHOVELERS c.10 PINTAIL 10 GADWALLS 2 *TUFTED DUCKS c.50
*POCHARDS 200+ MUTE SWANS SN *WHOOPER SWANS 60-70 Many juveniles
*BEWICK'S SWANS 200-250 (fewer juveniles) KESTREL 1
Welney – Littleport : c.48 BEWICK'S SWANS in a field.
The Wildfowl Trust observation hide at Welney certainly provides an exceptional opportunity to see wildfowl – all wild birds – at very close quarters. It was fascinating to see Coots, Mallards, Tufteds, Pochards and all three swans as close as 15 yards. I was able to see the details of vermiculations, the bright red eye and the droplets of rain on the Pochards; the green gloss on the cheeks of Tufteds; the bumps and depressions on the bare yellow area on the bills of the wild swans; the yellowish scaly pattern on the legs of young Whoopers and the brown colour of the irides on adult Whoopers – as this helps to indicate, they really were some views! To see wild swans of both species, in particular, as well as if they were in the hand was well worth the trip. In addition there was constant activity and musical calling.

We did the stretch north of Welney. You can see the results as recorded in my logbook. It remains an extraordinary day (but an ordinary one if you live nearby), briefly encapsulated in the notes with those remarkable numbers listed, including 9,670 wigeon, 330 pochards, 734 Bewick's swans, 75 whooper swans (notice only 29 mutes, however), a hen harrier, four short-eared owls. And look at the 50 tree sparrows!

Even so, numbers of wildfowl and waders were actually very low, as so much of the water was frozen. Forty-odd cormorants did what cormorants do, on the Washes, not that you would know from the average bird guide: they lined up together to roost on overhead electricity wires. Each time a bird arrived or left, the others balanced like tightrope walkers, or displayed with head and tail raised. Like this, against the sky, they are visible a couple of miles off.

Wigeon numbers were far below average – but for me, almost 10,000 was amazing, and didn't they look great. The light brightened enough near the WWT Welney reserve to spark brilliant crimson in the eyes of the drake pochards. At that time, I was using 7× binoculars, but the birds were close enough to show such detail as the eyes and all the 'vermiculations' (fine, wavy barring) on the flanks of teal and mallard. Of 637 Bewick's swans by the hides, 40 were juveniles, indicating a poor breeding season. The wildfowl counts can often reveal the breeding success of swans and geese, but not usually of the ducks, as first-winters (six months old) are too difficult to be sure of at a distance. They look just like the adult females before males begin to develop patches of adult colour. If you see, for example, a flock of swans of any species, or maybe brent geese close up at the coast (juveniles will have black necks with no white, but bars of white across the wings) you can do your own counts.

A water rail was so disconcerted after being flushed by some unknown disturbance that it skittered into a hide while we were inside, and left by one of the open flaps. After so many years, I had long forgotten this – but the notes bring it back. Water rails will kill small birds in harsh weather: this one may or may not have been the culprit, but once it left the hide it crept down the bank and began feeding on the wing of a dead bird. Finally, as so often happens here, the views of short-eared owls were exceptional: best for years, my notes remind me. They have such intense, piercing yellow eyes set in patches of black.

January is a good month for ducks and geese, and the WWT hides at several reserves offer opportunities to get to know the swans. Mute swans are always here, the ones that hiss and threaten and come begging for bread, but Bewick's and whooper swans are truly wild visitors from the far north and east, across into Siberia. Mute swans aren't mute, by any means, but lack the ringing or bugling calls of the others: on this freezing, misty day on the Washes, the bubbling calls of Bewick's swans were hanging in the air all day long, raising the atmosphere of the Washes way above the usual fenland land- and soundscapes.

Nevertheless, I remember well my first Bewick's swans, which I bumped into one November day not far from home on my local patch in Staffordshire. They drifted off on calm lead-grey water and looked at me, unsure whether or not to fly into the fog, as I tried to get as close as I dared but not too close, quietly bubbling their distinctively musical, conversational calls. Bewick's are not the same tame birds as mute swans, much more likely to take alarm and leave. Finding your own is *always* the best way, however good the views and overall experience from the reserve hides. Of course, seeing hundreds is wonderful, but they are *expected*: the surprise of a couple in an unexpected encounter, especially your first, is more memorable and so satisfying.

What is it, exactly, that can, many years later, recall such emotionally charged moments so vividly to mind? I can't really remember my birthdays at all, but my first green woodpecker, my first jay, are firmly embedded – and with some of these things the place is equally important.

***

The Somerset Levels is not an area I know very well, but one winter visit was rewarding, to say the least. It is an area that vies with the Ouse Washes as a premier wildfowl site.

West Sedgemoor, Somerset Levels

**Wigeon	10,000+ – remarkably dense, relaxed flocks
*Teal	c. 5,000 – constant noise, like a distant estuary full of waders
*Pintails	500–1,000
*Shovelers	500–1,000
*Gadwalls	200–300
Mallards	c. 20
*Peregrine	1 adult female – good views
*Merlin	1 female – fast, low fly-by, briefly perched
*Hen harrier	1 female
Buzzards	2–3
*Lapwings	5,000–10,000 – very widespread
*Golden plovers	c. 10,000 – distant
Snipe	1
Black-tailed godwits	30+
Dunlins	30+
Ravens	2+
Fieldfares	50+
Redwings	SN
Not a single gull!	

Tremendous numbers of birds: more densely packed in general than, say, on the Ouse Washes, but none particularly close. Great views, nevertheless, with a telescope. I'm not sure I've ever seen so many teal, and almost certainly never so many shovelers.

(SN and LN in my notes refer to Small Numbers and Large Numbers, with no very specific parameters – they depend on the context, so LN of sparrows in the garden might be 25, whereas 25 dunlins on an estuary would be SN. These were simply notes for myself and clearly are now of no real value, at a distance of many years, except to say that the species was present. * means something unusual or worth highlighting for unspecified reasons, such as a particularly good view, or an unusually high number, or a bird not often seen here, or similar, to lift it above the routine list.)

Greylake, near Glastonbury

Little egret	1
Buzzards	c. 3

*Wigeon	c. 1,000, some close – great views
*Teal	100+ – close views, really good – males brilliant
Gadwalls	SN
Mallards	SN
Shovelers	SN
Fieldfares	SN

Ham Wall, near Glastonbury

Little egret	1
Gadwalls	c. 50
Buzzards	6–8
*Lapwings	several thousand
Cetti's warblers	2 heard, 1 singing, 1 giving several loud *plip plip* notes
**Starlings	around 1 million perhaps
Lesser redpolls	c. 5

Starlings – vast flocks coming in to roost, but settling quite a long way off, far beyond audible range. Many tens of thousands appeared, then swarms more arrived (distant) and I thought I could easily believe, say 200,000 or more – then maybe 50,000, then 50,000 more, then more big flocks – swarms coming up like queleas from the reeds – so, I thought, there could easily be 500,000. Then another 50,000, another 100,000 and so on – if there were 500,000 before, there must, then, I suppose, be 600,000, 700,000 or more – and so it went on. So, I could believe a million maybe, without really making any great effort at trying to estimate them.

I haven't seen starlings like this for decades, perhaps never so many before. I didn't get the 'full performance', though – they rose *en masse* but not far and only a couple of smallish groups swirled and swayed around in dramatic shapes, briefly.

I have seen starling performances at other times, but not in such numbers. Even so, 10,000, 20,000 or 50,000 will produce dramatic, scintillating effects as they rise and fall, swerve and twist, funnel down and spiral back, all as if magically coordinated and synchronised in full flow. They are matched only, perhaps, by estuarine flocks of knots. More often you may see, say, 20,000 in a long, dense flock crossing the horizon, or pre-roost gatherings, maybe just a few thousand rushing by low down, really quick, their wings like a waterfall of noise. If they settle, their voices take over in a constant rushing babble. For years I thought 'murmuration' referred to this noise, but it is now used to describe the

Having fed together on the ground, the sharp-winged golden plovers (four now in summer plumage) separate from the blunt, sparkling lapwings in the air. Flocks of both are now much less frequent even in areas where they had been found year after year for decades.

aerial performances of the flocks: rightly or wrongly, I'm not sure. By the way, the queleas mentioned in my notes above are small finch-like African birds, sometimes occurring in flocks of millions. In Zimbabwe, they used to poison them wholesale and sell them in tins for food, poison and all – perhaps they still do.

MORE GOOD DAYS, BAD DAYS, ORDINARY DAYS

A day may be made special by the weather, companionship, or just good birds happening to 'click' all at once, or it may be a journey away from your usual area. Then 'your' good birds might be perfectly ordinary ones for someone who lives there. Scottish readers might get crested tits on the bird feeder; English readers see them once in a blue moon, if at all. But we can beat you for nuthatches! Ha: you'll lose a few if you go for independence. You will find memorable days popping up for all kinds of reasons.

Here are three successive days on a trip made from Swansea to East Anglia one February in the early 1970s, I think a minibus full of young birdwatchers, students with no money to speak of but just making do on a long and tiring journey. Where we ate, spent the nights and washed (if we did wash, I presume we must have) I can't recall. On other trips I camped, and floated out of the tent in the middle of the night (okay, I had a little blow-up airbed) as the torrents of rain gradually became impossible (one very wet night at Dunwich near Minsmere sticks in the mind). And I had nights on the floor in the George Hotel in Cley. I didn't ever resort, as some people did, to public toilets for the night, or kip under a hedge. Sometimes I slept in the car, or the back of a van (frozen to the bone: not recommended). A Holiday Inn or Travelodge would have been luxurious beyond our means. Anyway, none of us was then particularly familiar with Suffolk or Norfolk, so this was, however we managed it, a bit of an adventure in a region renowned for its birds and nature reserves.

Minsmere was the first and most famous location we visited: one of the all-time great RSPB nature reserves, on the Suffolk coast, not far from the Sizewell nuclear power stations, where Sizewell 3 will cast a long shadow. This was where the famous/infamous Bert Axell created new habitats and oversaw the reserve where bitterns, avocets and marsh harriers bred. Bert was supposedly fearsome, but then you shouldn't have trespassed. Many years later I drove him around and introduced him as a celebrity speaker at bird clubs and RSPB groups: a thoroughly nice man! Unfortunately, I didn't really get the time or the right circumstances to delve into his memories of life at Minsmere or earlier at Dungeness, which would have been fascinating.

A new bird! Well... not the most exciting, but still, I had not seen a wild Egyptian goose before. So it got the treatment in my notes. Egyptian geese were originally released in Norfolk, and at the time I'm describing were still very local, so you just had to go that way to see one. Now they have for some uncertain reason become much more numerous and widespread and people worry about the damage they are doing. You can almost walk up to them in some places.

Anyway, it was good to see some proper geese – not what Minsmere was noted for, but seven white-fronted geese, not to be sneezed at. Despite the date, two of the more noted Minsmere specials were still around. There was an avocet, always a great bird wherever and whenever, and two excellent marsh harriers. I'm pleased to see that I noted them down as females or immatures: I would not have been able to separate these really and, while many people seem confident in 'calling' a juvenile or an adult female, the distinctions are not easy. But not only were there two marsh harriers, but two hen harriers as well, this time an adult male and a female or juvenile.

Unexpected were two hooded crows. I was used to seeing hooded crows in Scotland but they do not breed in England and these would be migrants from the near continent or Scandinavia. Hooded crows used to be much more regular and numerous in East Anglia in winter, but, oddly, have declined recently, so these added a little extra to the day.

The Minsmere 'scrape' (an artificially created area of shallow pools, sand, mud and shingly islands, Bert Axell's pride and joy) had 75 or so twites, but they were a bit distant. But dotted about the scrape were other small birds: there were a number of larks, and then a larger group of larks of some kind flew up and settled closer, on the shingle beach. I counted 49 – and these were not just skylarks, but *shorelarks*: my first. They went back to the scrape, where there were probably more. This was pretty good, even for Minsmere. Numbers have since declined. (You might prefer shore lark, which means the name appears in an index as Lark, shore: but then, do you do wood lark and sky lark? If not, there's not much point.)

While they were on the beach I was able to lie back and watch with my telescope (we didn't use tripods, then, so lay down and balanced the 'scope on our knees), so I could see all the details of the head patterns and so on. Some males were 'perfect' except for the lack of the little 'horns' on the crown. My telescope was a bit short for this kind of thing: other people still used the older leather, brass and glass Broadhurst Clarkson drawtube type, long enough almost to reach their ankles.

We quickly moved on. Just up the coast is another large 'inlet' of reedy marsh with small areas of open water, at Walberswick. This little diversion proved to be well worthwhile. Six species of bird of prey included the expected kestrel and sparrowhawk, plus two buzzards (at that time, in the early 1970s, buzzards were really not often seen in eastern England), a red kite (a winter visitor from the continent, well before reintroduction schemes had established them in England – they were still very rare and restricted to Wales as breeders), a hen harrier and – another new one for me – a rough-legged buzzard.

There may have been two kites. They were familiar to us from Wales, after treks up to the central part of the country, but scarce enough here, and still 'good enough' for me to include descriptive notes, including plumage details, wing shapes and the characteristic tail-twisting actions.

One was right next to the rough-legged buzzard; at times, three or four large birds of prey soared together but, at a range of a mile or so, it was hard to be sure what was what. The rough-legged was great. Longer-winged than a common buzzard, it soared on flattish wings and often hovered, with a lighter action than a buzzard, using shallow, rather quivering wingbeats. Buzzards hover, too, so this is not a 100% identification feature by any means, just a clue, but rough-leggeds do hover more often. Fortunately, despite disappointing views, the wing shape and actions and the tail pattern (white with a broad, sharp black band) saved it. In the years since, I have still seen remarkably few rough-legged buzzards, although at least I have had better views.

On top of this, there were four barn owls hunting over the marsh in the evening; they are always 'top birds', and these gave lovely close views. A kingfisher, a couple of snow buntings (rare for us from the Midlands and west) and, so soon after my first ever, another 22 shorelarks, this time watched from about 10 yards. They often are surprisingly approachable. One of the snow buntings was, my notes say, 'the best I've ever seen', that is, a near full-breeding-plumage male, with the merest smudge of tawny each side of the chest, a hint of charcoal grey on the back and a few buff edges still to wear off on the rump. The black-and-white summer effect is achieved by wear and bleaching of fresh, tawny-brown feather edges, but the bill also changes, and this one was still yellow, not the 'summer' black just yet.

This was going well, then. Next day we were in Norfolk (where *did* we stay?) and visited one of the areas that has an unrivalled reputation in England. Every birdwatcher must some day go to Cley. Cley-next-the-sea and Salthouse combine to make a wonderful patch of marshy fields, reeds, open water, shingle and beach, with long, grassy, elevated banks and even a few bushes for migrants on little hillocks inland (Walsey Hills). I don't know what the figures are now, but in its day Cley had probably the largest bird list of any English parish: perhaps it still has. It was also the home of field-guide illustrator Richard Richardson.

Cley was not so productive as it might have been, except for one long-standing winter regular, an adult glaucous gull. Probably a male, he was a really huge one, much bigger than nearby herring gulls; given my interest in the species, more notes went into the book. The beach, though, did come up trumps with more snow buntings and again a handful of shorelarks.

It was at Blakeney that we got to grips (a typical birdwatching phrase back then) with an East Anglian speciality that we wanted to watch, a decent flock of twites. And Wells gave us a hooded crow, Holkham more Egyptian geese and a flock of golden plovers. Heading west, we ended up at Snettisham on the corner of the Wash. There are now more hides, a photographer's hide and the like, but then it was a bit bleak and we just tried to keep down low and inconspicuous. Whooper swans! Good. And looking down from a nearby clifftop, we could watch 52 scaup: the much scarcer marine equivalent of the freshwater tufted duck. Waders were not especially good by local standards: knots numbered maybe 5,000, bar-tailed godwits 1,500. We could have seen more knots back 'home' on the Burry Inlet north of Gower, where thousands would demonstrate the oft-quoted resemblance to clouds of smoke as they swirled and dived over the estuary mud.

By a forgotten route next day, we descended on Virginia Water, a place known to some of the party (not, I think now, to me) as a good one for hawfinches. I had never seen a hawfinch. It is a scarce bird, peculiarly restricted to smallish areas with particular sites known to be good year after year. You more or less go to one and hope, or very rarely just bump into one or two by accident somewhere. But the lake and lakeside woods had other goodies too. Mandarins! Brilliantly exotic-looking ducks brought across from eastern Asia long ago. And a red-crested pochard! I was less excited by that as I just assumed it was a feral bird, 'gone wild' after escaping from somewhere. Like the mandarins.

And in the woods, with marsh and long-tailed tits, a bonus in the tiny shape of a lesser spotted woodpecker. Lesser spots were never very common so always worth an asterisk in my notes, and they have since become worthy of two at least, if you can find one at all. But sure enough, there they were: hawfinches.

We had hoped for a hawfinch or two flying off over the treetops, struggling to be 100% sure that we could 'count' them. Instead, I saw one flying towards me and into a tree, then two or three more, before they dropped to the ground, began to feed, and were joined by others, making at least 18. Damn, just at the wrong moment, a car drives too close and they go: but no, only to the nearest treetop, before they return to feed. I watched from 75 yards with my telescope and from 25 yards or so beneath birds in the treetops, so views were very much longer and better than expected. Other birdwatchers got closer and dog walkers passed by between us, but they didn't go.

Hawfinches can be notoriously difficult, shy and elusive, but if you get a good view, you might be really lucky and see all the detail you need. These were performing to perfection, slipping down from the treetops a few feet at a time, dropping to the ground, fanning their wings and tails to reveal the patterns and peculiar feather shapes, calling quite often: the lot.

We still had a bit more to come, as we managed three black-necked grebes, some goosanders, surprisingly a red-breasted merganser and then 10 smews on Staines reservoirs. The west London reservoirs close to Heathrow were noted for their black-necked grebes and smews, which can be hard to find elsewhere. We went to look specifically, and there they were. For the three days we had more or less targeted local specialities and found them, so we had what must go down as special days, although for the locals, I suppose, they would have been ordinary enough.

It depends where you are from, where you go, and you just need a bit of local information, a bit of persistence, and a fair dose of luck. If you are going for 'regular' birds that are more or less guaranteed, it should work. If you are on a trip to see a rarity or two, it can be a miserable failure. I've missed four major rarities at different localities in one day before now. You get used to it.

New Grounds, old geese

Days that can't be repeated sadly include many at specially protected places, as well as the low-lying valleys of mid-Wales. The New Grounds at Slimbridge on the Severn estuary was a great place for geese, explored by Peter Scott when he was looking for somewhere to set up his reserve and wildfowl collection, and eventually to create the Wildfowl Trust (now Wildfowl & Wetlands Trust). In particular, it had a regular winter flock of white-fronted geese, often over 7,000. By the 1960s there were public hides from which the geese could be watched, with a bit of luck: they still flew around to feed quite widely in the area, so it was a matter of luck which field they were in on any given visit.

On 1 February 1970, things went our way. I had almost no experience of wild geese at that time. White-fronts in view numbered around 4,000 but only 300 settled close enough and long enough to be scrutinised. The blackboards said there were 7,600 that year. They are 'grey geese' (like greylags, pink-footed and bean geese), more brown than grey but more or less grey on the wing. Of these, the white-fronted is the most strikingly marked and most colourful, and in some ways perhaps the most agile and attractive (although pink-feet are perhaps more elegant in shape and were, I believe, the favourite geese of the Scotts). Adults have a white forehead 'blaze' and intensely black belly patches, and their bright orange legs add a splash of vivid colour against rich green grass. The 'yodel' call with a bit of a catch in it is especially appealing, too, creating a lively and energetic chorus.

If you have a big flock of geese like that, giving a decent view, it is worth scanning through them, as you would ducks on a lake, waders on the shore, or gulls at a roost. Start at one end, work your way through them one by one. Then go back and do it again. But it was when someone pointed a bird out to me and asked (suggested) if it might be a bean goose that I first saw it, my first ever: big, brown, long-headed, long-necked, with a dark bill crossed by a band of orange, and rather yellowish-orange legs. Although big and 'heavy', it was also particularly graceful.

Easier was a dark-bellied brent goose (more usual on the shore, occasionally 'caught up' in other goose flocks – easy to miss because of its smaller size in a flock like this). And most obvious, two barnacle geese: properly grey, barred with black and white, with a gleaming white belly, black 'breastplate' and neck and a broad white face. Barnacles! Exciting stuff.

Slimbridge had, by careful protection and judicious feeding, built up its regular winter flock of Bewick's swans (and, famously, Peter Scott and his daughter Dafila had studied them, drawing their individually identifiable face patterns – left profile, right profile, face-on – to help identify returning birds year after year). There were 430 and, from the high Holden Tower hide, we watched them fly out in full cry to the estuary, a very fine sight. Suddenly, as the swans settled, the sky was full of geese: what a noise! They came right overhead making a terrific din, but settled far away, out of range for identifying anything at all.

There were several other visits to Slimbridge over the years. One day in February 1972 was a good one, with 3,000 white-fronts well placed but a bit against the light; nevertheless, a fine adult lesser white-fronted goose was picked out. This north European rarity occasionally gets caught up in a flock of other geese and comes too far west in winter, to be seen in Britain as a rare vagrant. More recently, records have been slightly compromised, perhaps, by the fact that most are reintroduced birds from Scandinavia, a bit like our English kites. I had much better views in 1974, when the yellow eye-ring and short, vivid, almost shocking-pink bill was more clearly evident in good sunshine. There was often a lesser present there, and the problem was simply locating it amongst all the others.

Reedbed riches

Reedbeds are naturally restricted, developing in very shallow water undisturbed by strong currents, so they are found in coastal lagoons, at the edges of flooded pits, beside quiet rivers and so on. All are worth preserving and worth looking at. My experience with extensive reedbeds was mostly at places such as Oxwich on Gower, West Glamorgan; we didn't really have them where I grew up in Staffordshire.

In May 1978 I looked at several reedbeds and had some exceptional days.

It was a full-on, shameless expedition to see a rarity: I had never been to Stodmarsh in Kent before (nor, sadly, have I been again), but there was a pallid swift there. The chances of seeing a rare swift seemed remote. After all, swifts belt around in the sky all day long and can be anywhere an hour or two after they are first seen. There was nothing to tie it down. Yet this one seemed to be hanging around, with good feeding over the reedbeds. Off we went.

It proved a good decision. Five garganeys were great for a start. These are unusual ducks, in that they are summer visitors while most other ducks are more numerous in winter. Most of my garganeys have been autumn migrants, and spring birds with males in full plumage are a treat. Noted but with no particular emphasis were a few turtle doves and five cuckoos: both would be star birds these days. Half a dozen Cetti's warblers were singing, this at a time when they were still scarce and restricted in range, so another bird that was worth a bit of a trip on its own. I made a lot of detailed notes on them, as I had until that time seen rather few.

Bearded tits were usually seen on my occasional trips to the East Anglian coast but nowhere else, so about 30 of those added further richness to the day. I noted that these were more easily seen and obvious than usual and gave the best views

I'd had for years. Thirty, looking back at the lists, now seems a goodly number. Either way, they are lovely and fascinating birds.

Many reed and sedge warblers were singing, but more unexpected – and a new bird for me – a Savi's warbler added its low buzz. This is a warbler like a grasshopper warbler but without the streaks, more like a reed warbler to look at. It has the grasshopper warbler's trilling, insect-like song but with clear differences. Here are some of my notes:

> A very obliging bird indeed. Spent long periods in the morning and again late in the afternoon and evening singing, whilst in full view at 40–50 yards range. In an area near dense reedbed, but in very sparse reed stems above a layer of sedges. Apparently fed in the sedge but leaped up to perch on a tall reed or isolated reedmace to sing. Song immediately distinctive at close or long range, deeper, less shrill and ringing than grasshopper warbler's, notes *less* distinct, *faster* (not slower), producing much more of a regular loud *buzz* than a rattling trill. Presumably an unmated bird as song was regular and *very prolonged*, not in short bursts. I timed a single burst at five and a half minutes, followed by a break of a mere second or two. Other song bursts might have lasted ten minutes without a break...
>
> Bigger, longer-tailed and larger-headed than reed warbler with smaller, thinner, sharply pointed bill. Large head held up high and back when singing, bill wide open; tail angled down, giving an odd bend at the rump, slightly graduated with rounded tips often slightly separated, giving a ragged tip. Undertail coverts particularly broad and long. Small, curved wingtips well separated, revealing the plain rump.

The specific reference to the song being faster, not slower, is because one book had misleadingly put 'slower', which turned out, it is said, to be a misprint for 'lower'. The structural notes are quite important, as the rounded tail above very long undertail coverts and the markedly curved edge to the short wingtip are good features of this group of warblers, of the genus *Locustella*, compared with the reed-warbler types (*Acrocephalus*). There will be just a tiny handful of Savi's warblers (or even rarer lanceolated and river warblers) in Britain each year.

Yet the main aim was to see the rare swift. And we did. There were swifts about above the reeds, often going quite high as they usually do, all day, in their everyday astonishing way. Perhaps 300 scythed around the sky most of the time, enough to make picking out a different one a little tricky but not overwhelming. As it turned out, the pallid swift gave prolonged views, often at close range, so it turned out a bit easier than we might have imagined. It was seen in dull morning light and bright afternoon sunshine, so I noted that the plumage differences were distinct enough but not blatantly so, while the differences in structure from the other swifts were very hard to be sure of. That seems about right, following much more experience on foreign trips since.

The common swifts looked sharp-winged and slender, with slightly protruding heads, while the pallid had perhaps a fractionally less protruding head and slightly straighter, blunter wings and a broader, more fanned tail, but as the swifts would often spread their wings and tail in a tight turn, the differences could disappear. The frequency and length of glides varied greatly in both but the pallid tended, on average, to make longer glides with its wings markedly angled downwards. On the other hand, it often lifted its wings up in a V and tilted over to lose height, or stalled and twisted down like a falling leaf. On average it was more fluttery than a swift... but then again, both could do the very same things.

The plumage was more reliable, over the course of the day. It is interesting that recent reviews of their identification point out that single photographs can be very misleading. One bird in one photograph can even show features of both, or seem to do so: common swift in the shaded parts, pallid swift in the sunlit bits. To be sure of these, you really need to watch for as long as you can to get the overall impression right, and to see the tiny details such as faint paler barring underneath on the pallid. Or a series of photos from different angles.

But with one pallid all day long in a flock of common swifts, it was easy enough to pick out over and over again: a single bird flying by never to be seen again might be harder.

The very next day we were at another well-known reedbed, at Minsmere in Suffolk. The RSPB reserve there has a large number of breeding species in a great variety of habitat, but is most revered for its reeds and lagoons. As mentioned before, I had watched over the marsh and 'scrape' at Minsmere from public hides near the beach, but this was the first time I had walked around inside the reserve. It was all a bit overwhelming.

> It proved really marvellous. Though much attention naturally rested on the rarer birds, it was, at times, a job to know what to look at with so much going on. Many species were at very – almost incredibly – close range – in particular a drake gadwall, a pair of teal (showing every detail down to the barbs of the body feathering), shelducks, black-headed gulls (always calling, fighting, displaying) and several avocets. I was unable to fit a single avocet in the field of view – unbelievable.

Why keep notes like this? I had been watching birds for many years already and these things were mostly familiar enough, almost 'everyday' some of them... *of course* black-headed gulls would be noisy and squabbly at the colony, I knew that well enough. But I think I always wanted more than a list: a list can only give a little idea of what the day was like. I wanted to record the specially good things, why it was a better than average day: and above all, as I say elsewhere in the book, I think I always wanted, even if the species was common, to *do justice to the birds*, somehow. The teal looked amazing, the gulls were just brilliant (even though other people might barely have given them a second look) and they *deserved* to be mentioned. And an avocet more than filling the field of view would in any case have been something extra special. Avocets have a history, a special place in UK birdlife.

What I can't remember, as so often happens, is how I got there, where I stayed between these two days or whether (as I suppose I did) I went home in between.

But, as hinted at already, Minsmere was good for some rarities that day, too. Obviously, avocets were special: about 75, some on eggs, some with tiny chicks. There were top birds such as black-tailed godwits in breeding plumage, a wide variety of waders including good ones for a spring day such as wood sandpiper, greenshanks and two little stints, a little gull, kittiwakes, many common, Sandwich and little terns and a Minsmere special, the expected bearded tits.

Particularly special was a purple heron – only the second I had seen at the time, after an earlier memorable discovery by a group of us students at Oxwich on Gower. It was a brief view, only in flight, as it was being mobbed by other birds, stretching out its remarkably long, slender neck and dangling its big feet. Surprisingly, especially in view of their later great rarity in Britain, there were not one or two but *four* Kentish plovers, maybe more – all males. There were often three at a time, sometimes four together, even though in some years since, 10 *in a year* would become pretty good for the whole UK. All had the classic breeding plumage marks and the typical 'chick-like' shape and actions of Kentish plovers, which are like small, sparsely marked, rather bright ringed plovers with black bill and legs.

But things got better: a broad-billed sandpiper. Now I had seen a broad-billed sandpiper at Minsmere before... or at least I had recorded one. But it was distant and my notes now look unsatisfactory: did I see the real thing or not? I think so, but anyway, I could do with a better one, and here it was. Actually, in between times I had seen one at Breydon Water, and this Minsmere bird was a little less striking than that, but all the same, a very good breeding-plumage adult complete with all the necessary stars and stripes and markings that make it a distinctive little wader in that plumage. At 100 yards a pale V on the back was the most striking bit of a rather greyish bird with a bright white belly, and the head pattern was hard to work out. Close up, the head striping was in fact extremely striking, with a blackish crown edged by two broad white bands above a dark eye-stripe. The two white bands might appear 'forked' but in this case seemed just about fully separated, with just a hint of a dark 'join' at the very front.

On top of that, there was the unexpected extra bonus of another Savi's warbler, just a day after my first. It gave much shorter bursts of otherwise identical song from a little willow bush at the edge of the reedbed, and looked a little brighter.

Just to top off the day: later, I saw something that was still for me (and remains) a rare event, a magnificent breeding-plumage female dotterel. The significance is that the female of this species is brighter than the male, so this was as good as it gets. And then a golden pheasant: introduced, of course, but what a bird!

So not a bad day, all in all, the kind of thing that May can produce if you are lucky and if you move about a bit. Back home, I was content with a couple of spotted flycatchers, the local tree pipits and some kestrels.

More days that can't be repeated...

Sadly, many past experiences are no longer possible. This is one day in July 1977, in Suffolk:

> Golden orioles calling almost constantly for over half an hour but I only had the poorest of brief glimpses of movement once or twice, then a glimpse of yellow rump in dense foliage. Later in the evening, a glimpse, then one seen well enough just to make out black and bright yellow. Then I twice saw a male (or, at least, a bright yellow bird) fly across a railway in pursuit of a kestrel – very brief, but bright yellow body, long black wings with a square yellow patch and easy, graceful flight. Calls included a variety of fluty whistles, occasional low, piping, double call and hissing, squalling, catlike notes, like some soft notes of a jay, or even a squirrel.
>
> My most prolonged views of red-backed shrikes with those of the female 'best ever'. The male settled on a pile of felled branches in the centre of a small field and frequently flew out to catch insects. Often on a pure white birch stump in bright light against a dark green background, a real picture. Female superb, very close for long spells, flycatching, feeding newly fledged

> young etc. Called with chatting notes in answer to the whitethroat-like churring of the young.

I was fortunate enough to see red-backed shrikes a few times before they disappeared as breeding birds from England (they were once relatively widespread), but the golden orioles were best left for trips to Spain, where I did manage some great views. A male red-backed shrike is particularly appealing, with its soft blue-grey head crossed by an inky-black band, red-brown back, black-and-white tail and soft, wild rose-pink underside.

You can't see breeding orioles and shrikes in England now. And on the same day (with stone-curlews, too) turtle doves casually noted in 'small numbers' in five different locations. Just routine. Not any more...

Here's another day you can't repeat: in south Wales, 1970, at Dryslwyn/Golden Grove, in the beautifully rural Tywi valley.

> What a tremendous afternoon! Despite steady rain, we had incredible views of the geese, with three species in one field together, down to 100 yards range at times. There were 1,500 white-fronted geese, 16 pink-footed geese and 3 barnacle geese: both pink-footed and barnacle were 'new' for me at the time.
>
> From a mile or more away we saw the geese in wet, riverside fields, among many pools and floods, and we could hear them easily. Then we moved down and found them close to the road and had remarkable views. The white-fronts were superb, very colourful, with a large proportion of young birds. There was a large proportion of young birds. Their size varied considerably, as did their markings. Some had black only on the mid-belly, others were extensively marked over the belly and flanks, with a sharp line around the lower breast, which was clear, pale grey-brown. The legs were bright orange, bills pale pink.
>
> The barnacles were seen next, a pair and a singleton. The pair flew off but reappeared and flew in very close. Very pale, generally lavender-grey back, conspicuously and evenly barred, with beautiful black necks and chests and large whitish face patch. Long legs, and rather quicker movements on the ground than the white-fronts.
>
> Then I saw the pink-feet, noted by their partly black bills. Their legs were pink, mostly a very deep, brilliant pink, not a pale, washed-out colour as in most illustrations; their bills very small; heads dark brown, 'shoulders' and breast pale orange-pink-buffish brown; upperparts pale blue-grey with

> white and browner bars, underparts pale brown. As large as white-fronts, more rotund and generally paler but darker-headed.
>
> Very active, the birds fed, argued, threatened and flew about often, and all the time birds were calling a wide variety of beautiful, evocative calls.

These Tywi valley whitefronts, with pink bills, were of the European/Russian subspecies, which used to be more widespread in England, not the Greenland one that winters in west Wales around Ynys-hir and is more widely found in Ireland. Today, there are no white-fronted geese in the old Welsh sites such as the Tywi valley and by the Severn near Welshpool. Even at Slimbridge, the famous Severn-side site in Gloucestershire, few remain where Peter Scott used to watch thousands and people like him and Ian Wallace would find the occasional lesser white-front too. Luckily, I did go to Slimbridge when numbers were high and did manage to see a lesser white-front or two (and even Peter Scott). White-fronts are doing okay for now, but no longer bother to come so far west and remain all winter in places such as the Netherlands. I used to go to the Netherlands for winter long-weekend bird trips and see tens of thousands of geese, and it was undeniably brilliant, but it is sad that these great birds have been lost from Wales. There are not many birds that are very large and also in flocks of hundreds. Just watch the gulls instead...

And another unrepeatable experience

In 1973 my friend Peter Garvey organised a trip for us to Fetlar, Shetland. It was a bit of a slog: first I had to get to Bristol, then we travelled to Glasgow by bus, and on to Aberdeen by rail, then took the *St Clair* to Lerwick. By the time we arrived I was 'bouncing' even on dry land: too much travelling, too little sleep, not enough to drink. We straight away hopped on a boat and took a trip around Noss, looking up at the great gannet colony. For years later, as a kind of trophy, the inside of my camera case was stained with white gannet droppings. Then we went to Scalloway to see long-eared owls in suburban gardens, and we did. They don't have many trees to hide in anywhere else in Shetland.

Fetlar was astonishing: a revelation for someone from Midland England, even with a bit of mainland Scotland experience. Shetland landscape is a bit Marmite, you love it or you don't; a bit of an acquired taste, perhaps, or just something that immediately appeals. To me, the bleakness, the wildness, perhaps to some extent the harshness, the unspoilt nature of the whole place seemed perfect. Great and arctic skuas were everywhere; there were whimbrels, golden plovers, lapwings, curlews, redshanks, snipe... flocks of *real* rock doves (genuinely handsome birds), groups of twites. Red-throated divers called while a great northern in full breeding plumage sailed on the bay opposite our little hired cottage at Tresta. We could read our books at midnight with no lights. Offshore were Manx shearwaters; on the cliffs were auks and fulmars and kittiwakes. We came across a merlin's nest with eggs and went to see red-necked phalaropes on little lochans. These were fabulous little birds, combining great rarity (in the UK) with real appeal, feeding within a few feet.

But the main aim was to see the snowy owls that were at that time breeding on a high Fetlar ridge, Stackaberg. Bobby Tulloch had found them and the RSPB set up a basic viewpoint hide to control potential disturbance and allow people to enjoy the birds from a safe distance. We wandered up to the hill, pausing briefly to speak to the woman who rented us the cottage. If I remember right, she had not left the island all her life. We soon reached the hide, and quickly saw the pair: the male, a 'little snowman', perched at 300 yards, the female on the nest about 120 yards off. There were in fact *two* nesting females then and this was the younger one, more barred than the other. The female became alert and raised her head just as the male flew, briefly. Later we watched the female stand and scratch, with hugely feathered white feet, moving away and then flying back to the nest, settling with a seesawing action of the body onto the hidden eggs. The male was sparkling, his yellow eyes looking surprisingly dark in contrast with the white plumage.

We watched the female again on other occasions, but then came a special day. There was an immature owl on Stackaberg, at first thought to be a female as the warden was at that time checking the nests and the adult females might well have flown a little way off. However, as it turned out, both adult females were by then back on their eggs and the male flew in, too. A female flew off, the male moved to another perch, giving marvellous views. Later, there was the adult male (there was only one, paired with both females) over the ridge with a female or immature, and remarkably two more appeared in the air as well: *five* in view at once! It was unexpected and incredible. 'Don't it get complicated,' observed the warden, Cliff Carson. The latter two seemed to be an adult female and an immature, and they flew together, grappling with their feet in mid-air, spinning and falling to the ground. One flew away while the other was mobbed by eight arctic skuas. Common gulls and arctic terns also mobbed the snowy owls, which reacted by jerking the head up and back, bill wide open, as the aggressor dived in. There were several calls, nasal, wheezy screams like *kee-airrrr* from the owls.

This was even better than we could have expected, and Shetland delivered to the full: storm petrels, the great gannet colony of Noss, long-eared owls... wonderful. There was a black-browed albatross at Hermaness, but we (rightly) decided that it would waste two days getting there and back, and the chances of seeing it were slim, so enjoying Fetlar to the full while we could inevitably took precedence.

Shetland remains extraordinary, but seabirds, including arctic skuas, are in serious trouble, their numbers spiralling downwards like a plunging gannet: and the snowy owls long ago ceased to breed. So that special day with five snowies really cannot be repeated.

WAXWINGS

A good breeding season for waxwings in northern Europe, combined with a poor berry crop for whatever reason in the autumn, sees hundreds or thousands arriving in October, November or December on the east coast of England. In a really good year, which does not happen very often, they spread widely across the country, stripping berry crops as they go, and even people in the west of Britain can reasonably hope to see a few until the spring. But finding your own will be unusual, and most of us 'go to see' waxwings that have been reported somewhere, whether it be a large old hedge full of berries, a park with small, ornamental apples still on the trees, or a woodland area with berries in abundance. Later on, they may turn to new shoots on trees such as poplars, or catch flying insects, reminding us that in the north, in the breeding season, they are accomplished flycatchers. Having said that, you can find your own simply by chance, maybe while driving along a country lane, as they tend to sit out on top of a tree or bush, distinctively silhouetted.

As berries and similar small fruits are so important to them, waxwings are often found in urban or suburban areas, including parks, large gardens and borders of supermarket car parks... you never know. Given a diet of tough berries, waxwings are also great drinkers (as are crossbills, which eat dry seeds from cones) and like to come to the ground to take advantage of a pool or puddle. Whether at water or when feeding, they can be remarkably confiding, but the temptation to push them too hard – to get just one more even better photo for example – should be resisted.

Waxwings are extraordinary birds in Europe (strictly, Bohemian waxwings, to distinguish them from the cedar waxwing of the New World). They are starling-like in form, stocky, short-legged and short-tailed and upright when perched (but acrobatic, like little parrots, when feeding), but have long, tapered wingtips and a very short, stubby bill. The obvious feature that picks them out is a tall, wispy crest. Colours are subtle, mostly pinkish-brown, pale buffy-fawn or pinkish-buff, with grey and black on the tail, rufous around the face and darker reddish under the tail, an inky-black chin and wings marked variably with black, white and yellow. There may also be waxy red feather tips on the wing, which give them their name. The detail of the markings can help sort out young birds from adults, males from females and, while the shades are soft and subtle, the overall effect can be really pleasing.

From individuals you hear distinctive bubbling, silvery trills, quite shrill, like tiny lightweight metal balls rolled together, and from flocks these calls make a quite striking chorus. Both greenfinch and blue tit can give calls that are confusingly similar, though.

Seeing a waxwing might become unexceptional given enough years of birdwatching experience, but should always be a real treat. And for years early in your 'career', if you are anything like me, it would seem to be a highly unlikely event, just a dream. One day, there might be a waxwing. You see the pictures in the books, and wonder if they can really look that good. And of course, they do. All the same, as they tend to sit about doing nothing for ages, or simply feed on berries in front of you, watching them is generally easy enough, and not much really happens, most of the time. But they are great birds all the same, and some flocks seem to be particularly active and entertaining.

My first were just three in a small Staffordshire garden, close to Blithfield Reservoir north of Rugeley, followed by a very thin trickle of ones and twos, then half a dozen, mostly in hedgerows. It was years later that bigger irruptions brought me into contact with larger flocks, the kind you dream of. In some years I couldn't seem to miss: waxwings even flew by the house and sat in trees opposite my office during the day.

One January I wandered around Staffordshire and came across 18 – but at the same spot the day before there had been 360! And then, shortly afterwards at a different place, there were 63, a more than decent number even in a good year, then four 'at work' that increased to 23 before I went away for a couple of weeks. After I returned there were as many as 48 close by. Never mind the 360, these flocks were brilliant by any normal standards, with plenty of action, colour and variation to look at.

The 48 flew by over a cycle track to settle in a hedge, still with a few berries even as late as the end of March. Every so often they drank from a flat shed roof or a puddle on the ground. In the afternoon they reappeared from nowhere, fed from berries low down at very close range and then withdrew to the tops of some tall trees. They were apparently wasteful feeders, even with berries now in short supply after the long winter, dropping as many as they swallowed, but a few tried rose hips, which they swallowed with considerable difficulty. They called their fast, tinkling, shrill trills all the while, at different volumes. Two raised their crests, fluffed out the body feathers, especially the grey rump, and exchanged berries bill-to-bill, presumably courtship feeding – but both had a berry at the same time and they 'collided' beak-to-beak. It was peculiar that the bird doing most of the displaying had the dullest wing feathers of all, with just white primary streaks (no yellow) and no red waxy tips to the secondaries. Adult males would be expected to have a full set of red tips and yellow stripes on the wingtip.

In fact, there was a wide range of wingtip patterns in the flock: white streaks with no red; white streaks with red wax but no yellow; yellow streaks with red wax; yellow streaks with thin, curved white tips; and yellow streaks with broad yellow and white primary tips and vivid shiny red waxy secondary tips (but no red tips to the tail, which are usually illustrated, but maybe wear off by late March).

A few days later there were 10 from my office window plus at least 58, now, in the flock down the road that had numbered 48 before. These were now feeding in tall poplars and maples, so the birds were half-hidden (or even practically invisible) in a mixture of fresh green and crimson leaves, yellow flowers and large dark red bracts, against a vivid pale blue sky. The birds themselves looked particularly slender for a change, and while they fed fitfully on the flowers they called continuously.

Other years brought some really big flocks. One year, my best was a magnificent 163. As I arrived and peered around, there was no-one else about, nor any birds at first, until I chanced across 22 in some roadside bushes – brilliant, with nothing much to eat but calling all the while. The usual quiet chorus would periodically burst into spells of really loud, excited sound. Then a bigger flock appeared at a puddle of water before flying off – now 40 or more. There were males with the full wing pattern and long, red waxy tips, but no red tail spots. The whole lot disappeared but a flock then flew over the local Burger King, maybe including these 40-odd, maybe not – but 163 settling to become easily visible from the car park. In flight, they moved in a tight, highly synchronised group, fast and furious like a flock of small waders as much as starlings, but each with a typically stocky, long-bodied, arrow-shape.

This happened to be part of a great day with large flocks of snow buntings and groups of lesser redpolls, mealy redpolls and a few arctic redpolls: whether one, two or three species, you take your choice, but arctics are particularly rare. Then there were 10,000 pink-footed geese, a barn owl, even a little auk...

But a few days later, in the very same area where I saw the flock of 63 waxwings in a different year, there were at least 120, in a housing estate with little green spaces and ornamental fruit trees. They fed, called, then flew off in a long, trailing flock, dashing downwind at great speed. Many settled on roofs and television aerials, real suburban birds despite their northern forest origins, and 30 settled on a tiled roof beside a smoking chimney. Orange rowan and red cotoneaster berries provided the main sustenance here, as well as a scattering of catkins that had fallen to the ground: birds were calling, fluttering about, hovering, hanging upside down, all within 10 yards or so. Individual waxwings sitting still are excellent, but it is a busy flock such as this that really provides the entertainment.

SOME DUSK ENCOUNTERS

I have watched nightjars many times: sometimes in a roadside strip of plantation with high trees as song posts; sometimes from a forest track into a deep, undisturbed woodland; sometimes from the top of a ridge on Gower with the lighthouses of the Bristol Channel flashing in the far distance; sometimes on wide open heathland with scattered birch trees and rolling slopes of heather and bracken, with Venus and Mercury settling lower in the bright western sky. On occasions after a long break, I have remonstrated with myself in the notes: why have I not done this for so long? How can these not be instant favourite birds?

> At the top of a long bracken slope, about 10 p.m.: two nightjars flew by at a range of eight or ten yards, still with a clear sky and half moon: amazing view. Both dark brown, no detail discernible but on the male, white wing and tail-corner patches stood out extremely well. They flew off along the slope, the male with slow, irregular wingbeats and glides with wings held up in a 'V', then swung downslope, slipping around birch trees and finally going out of sight. A single soft *go-ek* note then the male was heard churring and I followed it up, crashing through the bracken far too loudly in the quiet of the darkening night, but I could hear it very well. While listening to this, and the barking of a deer nearby, I saw a female appear; it flew around, circling and sideslipping, dipping back and fore close to the birch bushes, coming within 10 feet! It occasionally hovered for short periods, with body and tail vertical, and perched briefly on a twig, then briefly on the ground. A male appeared, 30 yards off, settled on a branch and churred for four or five minutes. It once 'ran down' with a broken, clucking end to the long churr but usually it broke off rapidly, often coincident with a quick foray from its perch, before returning to one of three dead twigs on its small branch. At this range the churring was loud: with my hands cupped behind my ears I could hear a really remarkable hard, rapid, but fairly musical rattle, with shorter, higher bursts as if the bird was 'breathing in'. Later both male and female were around, flying within a few feet of me, with their superb easy, leisurely flight, usually quite silent but very occasionally with a slight 'whish'

of wings. Thy would glide, circle, hover, dive, sideslip and waver so lightly, so buoyantly, like giant moths (not altogether unlike storm petrels with their gentle, stroking wing action). The male hovered and called a loud, vibrant, whistling *goo-eek*. The sound of the churr is really extraordinary and matched only by a closeup corncrake in the dark for peculiarity.

...

Nightjars in the same area as before and another churring (and one seen in silhouette against the sunset) farther west.

The closer birds were a bit complicated. One began churring intermittently very early in bright light, then one seen brilliantly, briefly, in flight, when there was still a bright sky and a lot of light – a 'full colour' view. Then only intermittent churring and quite a lot of *gooik* calling, at least two churring close by plus the third to the west – then, at about the time when it might be expected nightjars would 'start up', there was quite a long quiet period. Then, with still some decent light, churring began again and we had a series of very close views of birds in flight near or over the broad track. Several times one or sometimes two settled on the track and one even churred from the ground in full view – when there were two close together, they were both males, while another churred nearby. So there were three males but I only definitely heard two churring, sometimes both together in 'stereo' – they were remarkably close together. Brilliant views, and no sign of a female.

Woodcock roding, in many different directions but often overhead – sharp calls easy to hear, croaking hardly at all – I could see the bill opening with the calls.

All under a beautiful blue sky with pastel grey, cream and pink clouds and a nearly-new moon; good chorus of song thrush, willow warblers and several yellowhammers, occasional tawny owls.

Late-summer evenings are often particularly memorable and evocative:

A magic night, like going back 10 years... braving the biting midges until 11 p.m. First a nightjar, a female, floated up to perch on the edge of a silver birch, close, showing its pattern quite well, especially a pale superciliary, pale moustache, dark band through the eye and a prominent pale wing-bar. Then a male appeared in flight, cavorting about, often very close, its white patches gleaming like lamps, especially the wing spots. From head-on the white in the wings made a blurred arc each side with the rapid beats

Nightjar, with Venus and Mercury slowly sinking to the right. (Even eminent authors such as Tolstoy make the mistake of saying the 'evening star' is rising in the west...)

between silent glides. It called repeatedly with a very short, sharp note and later the typical, louder, longer, curious *goo-ek*. He perched and the female flew around, 'bouncing' in jerky undulations over and around dense birches, then settling on our track. She settled with wings slightly open as if brooding, then moved a few yards and did the same. The bird did this many times and, judging that it might be some form of distraction display, we dutifully cleared out of the way. The male then, for the first time, churred. Shortly afterwards one or maybe two others were churring not far off. Then another male churred and several times finished with a remarkably long 'run down' for 10 or 15 seconds before the churr stopped altogether, during which time it clapped its wings. Using the usual technique, cupping hands behind the ears, these calls and churrs all sounded tremendously loud and impressive.

In the sky I could see Venus, Jupiter, Saturn and (faint but recently confirmed over several nights) Uranus, all together.

...

It was a dull, drab sort of evening at home, but out in the woods and clearings later it was magical. It was dead calm, midgy, with a clearing sky after a day of heavy showers and dark cloud. A thin sliver of moon emerged low in the west, soon to set. It became cool and a thin, white mist formed at the foot of the trees, visibly rose from the ground and crept out across the clearings. It rose half the height of the trees and dispersed, as a new layer formed on the ground below. Dark pines and spruces mixed with huge old oaks and limes. In the clearing rosebay willowherb is just beginning to bloom; already making a magnificent show are clumps of tall foxgloves, many of them beautifully pure white. Birds were singing everywhere, sounding rich, loud and far-carrying in the damp air. Many pheasants added to the chorus, along with cuckoos and blackbirds, but it was a couple of tree pipits and a brilliant chorus of song thrushes that dominated. There was plenty of action from woodcocks for well over an hour. One noted on previous evenings, with a distinctive broken right wingtip, was often seen, plus at least two undamaged ones. One was individually identifiable by its 'broken' voice, the normal croaks separated by extraordinary deep squeaks. Certain lines were regularly followed (by more than one bird, sometimes two together). Two would apparently fly amicably together; on one occasion they chased with rapid, twittering whistles. Individual routes were, nevertheless, very variable despite certain elements being rigidly repeated. Suddenly, well through the twilight period – I'm sure it was sudden although I was properly aware of it only a little while later – everything stopped and the world became totally silent.

While a grasshopper warbler was 'new' in this location, sadly past nightjars now seem to be completely absent.

Evening skies and nightjars at dusk are particularly appealing, with a special serenity or tranquillity about them, especially as time is short and it will soon be dark, but I often stare at the skies by day, too, from the rare days of total, unsullied blue to dramatic effects with enormous clouds. I'd sometimes like to perch on top of one.

If the Fens are at their best on a day of big, dramatic skies, then this was it. After a lot of rain, the clouds began to break a little, finally to produce a lot of sun, but there remained huge, awesome clouds and occasionally a great, black, threatening thundercloud would rumble over. The enormous thunderheads were huge, broad, softly blurred and wispy-topped. Others were massive banks of more solid-looking, compact cloud with more intricate and complex shapes, with great chasms and rifts between them – vast mountain ranges cruising through the sky, themselves wreathed

> with thinner cloud, draped like a cloak over their shoulders. Imagine the thunder-gods living up there, striding about and sparring (like Tolkien's stone-giants!). Then rows and rows of smaller cumulus drifted across, miles and miles of open, flat land evenly covered with an armada of clouds. If clouds were not there naturally and an everyday occurrence, who would ever imagine such a fantastic concept?

I don't think we get as many thunderstorms as we used to. These caught my imagination and, unusually, my notebook, in 1983:

> 1 June. In the early hours and through what passed for dawn, there was a ferocious thunderstorm – one of the most continuous electrical storms I can remember. As the sky slowly brightened it became a peculiar green-grey, but every few seconds (usually 3–5) there would be a lurid flash of magenta. The whole sky momentarily flushed as the discharge flickered across the sky. Rain was usually nothing special but for a time became extremely heavy, but it was the fierceness and regularity of the flashes, the amazingly wide 'cover' of each, that made the storm such a spectacle – each sharp flash created a flickering blanket cover over the whole sky, sometimes leaving a persistent pale pink flush persisting between bolts, with a whole range of thunder from a short, sharp crack, through crashing bangs and strange, extended hissing noises to distant rumbles and deep, booming rolls.
>
> As soon as it moved away, a song thrush began to sing.
>
> ...
>
> 6 July. Remarkably dark all of a sudden by 8.30 p.m., with threatening green cloud and mist. The electric went off immediately and for 30 minutes there was a violent storm, then another half hour of rain and flickering pink and orange flashes. The peak of the lightning was tremendous, awesome, if not quite so continuous as on 1 June. Several long forks of lightning clearly came to earth and one thick, zigzag band of pink seemed to pour itself into a tree for several seconds. Another real 'thunderbolt' crashed down beside an isolated house, ending in what looked like a rather tame shower of yellow sparks.
>
> Once the rain subsided, a blackbird flew to the rooftop and began to sing.

SOARING BIRDS AND BIRDS OF PREY

Although I have once taken control of a small Cessna for a short while, doing a few simple twists and turns over Bridgnorth (okay, I should have warned you first, sorry), and have even experienced a hot air balloon flight, my only attempt to sail around like a buzzard was short-lived. The 0–60 acceleration in under 4 seconds behind the tow-rope as our glider took off was something of an eye-opener (even a knee-trembler), but the pilot soon looked round despondently and asked how much I weighed. Okay, we'd perhaps better go back down. But I did also fly a couple of times over south Wales, in another small Cessna but this time in the back seat, with a pilot who was also an expert glider and floated his plane around a bit, sometimes clipping the top of a little cloud and sliding down the other side, so I got a little bit of the general idea of what it might be like to be a bird...

July. Sitting outside my back door, with these lines going through my head; hot sun, cloudless, no wind, lifeless, listless, heavy, nothing to exploit, not a bird in the sky. Earlier, it was hot but with puffy little cumulus, what I think of as an 'Impressionist' sky, a bit of a breeze. Some movement for birds to use: more life, more uplift. There was a red kite, all angles and curves and long, slender wings, then a grey heron, broad, heavy, deeply bowed; then the next soaring bird was a distant black dot that surprisingly turned out to be a cormorant. Surprising for my garden, not for the cormorant: they *do* soar, sometimes, and quite well, too. It was circling high but had to flap quickly every 10 seconds: not like a buzzard or a gull.

Finally, a buzzard appeared, as good as it gets: soaring for minutes on end without so much as a twitch or flex of the wings. On a day with small rounded cumulus gathering into bigger stuff, a buzzard can be just above you one minute and hundreds of feet up the next. And buzzards could take me back to Wales as a schoolboy, or Scotland: I didn't see them anywhere else, then. They are too frequent, now, to be a trigger for nostalgic daydreams every time; but they remain splendid.

I can't deny that it would be good to be somewhere watching the vultures. I was disappointed to see Britain's first wandering young bearded vulture (I still call them lammergeiers) described as 'lumbering'. Perhaps it was, on a cold, wet, drab day somewhere in flat middle England, but given a bit of sunshine or a light breeze over a dramatic Pyrenean gorge, this is one of the most gloriously perfect flying machines there is. A lammergeier is big and heavy, sure, but it soars and glides with a magnificently perfect, delicate touch, if anything even a mite more extraordinary than the griffon vultures it often associates with. A glider pilot in the Pyrenees told me how he would often circle with the griffons, and how high they go. I may have written it down, or just assumed I would remember, but I've forgotten. Up to the cloud base anyway. The similar Rüppell's griffon has been recorded at 37,000 feet above sea level, but how far above actual ground level if they were over mountains? I'm sure griffons go well over 10,000 feet, and Seton Gordon reckoned golden eagles went that high above mountain peaks in Scotland. Oh, to be in Spain, or Crete.

Why do these large soaring birds make such an impression? Not just birds of prey, either, but storks and pelicans, if you like, although the storks are marginally less impressive in just one way, as their flocks swirl and crisscross at random, while the soaring pelicans move in synchronised harmony. And soaring gliders (sailplanes) up around or under a towering cumulus cloud. There's just something about being high in the sky and freed from the effects of gravity. But birds of prey have a charisma, an aura, about them that is difficult to explain and not always logical. From my back garden I can see the occasional sparrowhawk and frequently a buzzard or two (although in recent years these have unaccountably declined – at one time six or even nine together would not have been so unusual). Most days, if I try hard, I might now see a red kite; and now and then something else, such as a peregrine or a goshawk, even on special occasions a honey-buzzard. Recently a couple of hobbies. But ninety-nine times out of a hundred a high, large, soaring bird will be a rook or a gull.

And what's so bad about that? Most people don't even notice birds anyway (sad but true), and most birdwatchers wouldn't bother much with a soaring rook. But I think rooks are grossly underrated – and they can really fly. Quite why they go so high and circle round is questionable, as that seems to be of little or no benefit to a rook – not like the griffons that spill out of the Pyrenean foothills and use the height to glide away effortlessly, for miles over the adjacent Aragon plains, in search of a meal. But rooks do it for the fun of it (often accompanied by the slightly less accomplished jackdaws) and they can be quite acrobatic. Jackdaws will always get together in a flock, shoot out in a noisy, clattering panic and simply circle round to come back again, surely just enjoying the trip.

But for excellence in long-distance, high-altitude soaring and gliding, for minutes on end without a beat, the herring and lesser black-backed gulls take the biscuit.

Or the fish and chips. I'm a gull man, sure, but you can't deny that these birds with their long, slender wings, with a perfect aerodynamic aerofoil cross-section, are stunningly good fliers. And now and then they go astonishingly high.

So why feel a momentary disappointment if the distant high-flying speck turns out to be a gull? Well, of course, it is silly to pretend it wouldn't be better if it was a honey-buzzard. But purely from the point of view of admiring the flying skills, gulls are good enough.

It is interesting to think further along these lines. Peregrines have been over-hyped as the fastest animals on earth, flying at 200 mph or whatever the claims might be – but I can't help wondering how many people might be disappointed, having been attracted to a peregrine viewpoint, hoping or expecting to see something like a military flying display... and finding a peregrine sitting on a ledge, or soaring by at low speed.

Birds of prey, generally, fly *slowly*. In very many years, with many great views of peregrines, I can count the times I've seen a full-on top-speed stoop on the fingers of one hand. Now and then I have heard the great rush of a bird diving past. And I've seen goshawks in extraordinary headlong plummets from high altitude that must pretty nearly equal the speed of a stooping peregrine. People who know golden eagles say much the same about them. But peregrines often wander about the sky at a slow pace, or turn away in a more determined, direct flight with quick, deep beats of those long, tapered wings: not, perhaps, quite enough to catch a determined pigeon in a level pursuit. I've watched peregrines chase rock doves for half a mile or more and still not quite catch up. Many 'kills' are from below, taken in a precision roll with legs extended upwards. Yet, as my old friend Tim Cleeves said when he was employed as a peregrine warden in Wales, he would never take his binoculars off one while it was still in view. (Peregrines, to be fair, were rare then.)

Which brings us to pigeons. If you want speed, action, acrobatics, look for the feral pigeons around a seaside resort pier. They tip and tilt, soar and glide, dive with closed wings, swoop by with wingtips held up in a high 'V' and generally fly at a level above most competitors, often at fantastic speed. Wild rock doves around a north Scottish sea cliff do the same. Come to think of it... oh those beautiful northern cliffs and sea caves!

Here's a sleek, grey shape, dashing by, swooping at vast speed across a field, shooting up in an impossibly quick, abrupt turn, to settle in a tree – wow, *fantastic* – oh, it's just a woodpigeon. Oh well, next time a peregrine or a goshawk. But if the appearance on the scene was so fast and spectacular, what's so bad about a pigeon?

Okay, I would prefer the peregrine too. But enjoy what you've got. If my garden watch produced a kite or a falcon, all well and good, but the rooks and gulls are okay, too. Especially against a dramatic sky. A high, flying speck, be it a buzzard or a crow, can suddenly create a heightened perspective as it passes across a floating cloud: for a moment, there is extra depth and space, the sky reaches up and away to the end of the world, and just what the bird is that helps create the effect is of little consequence.

By the way, if you find it hard to separate a pigeon from a hawk or falcon at a distance in a brief glimpse, think about the local starlings: they are expert at it. Starlings fly off shouting at the first hint of a bird of prey half a mile or more away, yet never move if there is a distant silhouette of a pigeon. Or even a pigeon dashing in unannounced over the garden wall.

Birds of prey often display in spring. Not all will be in the UK in March, but some do display very early in the year. Forests have secretive and rare birds of prey, which *should be left entirely undisturbed*, but you might be able to watch inconspicuously from a secluded spot on a rise nearby, looking out over a valley or extensive wood.

Watching goshawks

Find a big wood, deciduous or conifer it doesn't much matter so long as there are big trees, with a viewpoint alongside where you can sit comfortably and look out across the treetops. Try to be inconspicuous so as not to attract attention if there are sensitive species about. Choose a fine, dry and still day in March, probably about mid-morning or midday, not necessarily very early. Keep still and quiet and watch. With luck a pair of sparrowhawks will be at home, and they might perform their territorial displays. If there are two, you will see how much bigger one of them is – that will be the female. One or two will fly over the wood, at a fair height, with a slow, steady pace and slower, deeper wingbeats than their normal snappy flap-and-glide action. The flight might even recall a harrier, or more likely prompt thoughts of the larger and much rarer goshawk. You could see the white undertail coverts widely spread, also like a goshawk. If the display progresses, one may suddenly begin a series of undulations that become deep, vigorous swoops, 'bouncing' up at the bottom of each dive with almost closed wings. They may soar, too, and finally disappear into the wood with a headlong plunge.

If you see this, or perhaps even a goshawk doing the same thing, you might wonder why the peregrine is always credited as the world's fastest bird: these hawks look every bit as quick, and simply disappear into the trees still going full-speed. Sparrowhawks may be elusive and low-level hunters much of the time, but they really are spectacular at times. Leave them alone (if you see goshawks, you would need a licence to do more anyway) and wish them luck.

Sparrowhawks are splendid birds to watch like this, rather than simply as a quick glimpse as one dashes by, but increasingly it is the bigger goshawk that becomes the focus for a forest visit. Goshawks are still rare, but with a decent population in some larger forests now, following a slightly obscure history of escapes and illegal introductions. How much has come from natural colonisation is uncertain, but probably little or nothing. Even so, it is a little different from, say, the red kite situation. Most kites are certainly from introduced stock: it doesn't minimise their appeal, but there is just a feeling perhaps that anyone can see a kite now, without trying. I recently watched several, over the busy central streets of Windsor, and I always do a mental count of birds seen on a journey:

My early goshawks in the UK were in the English Midlands, and some of the birds we saw then looked remarkably big: females like buzzards with long tails. Probably released and escaped goshawks were from several different origins, some larger, some smaller. Some of the bigger goshawks looked very pale, suggesting a northern origin; smaller birds in southern Europe can be much darker. Often adults look greyer above than might be expected; at long range, you can sometimes easily see the bright brown and buff of immatures, the pale buffy underside catching the light as they tilt over.

even from the M25. But goshawks take a bit of seeing, introduced or not, and it remains a bigger deal if you see one well.

March. Sunny late morning and midday period (before it would become very dull) – right time, right conditions, let's try for a goshawk – and success! Off to one of my best goshawk spots. Here you can see goshawks, sometimes honey-buzzards, sometimes hawfinches; it is a good area for woodlarks. Yet you can't take someone there to see any of them with a guarantee of success. These places and these birds don't work like that.

Buzzards were up high and also diving and switchbacking low down, with some fantastic choruses of loud, ringing, challenging calls – superb. I think it was Colin Tubbs who said the buzzard's call, so often called a 'mew', is sometimes such a dramatic one. Then I noticed an obvious goshawk and watched it for perhaps 20 minutes. It 'glowed' quite warm-coloured in the sun so, although it was hard to be sure, I think it was probably in buff, streaked immature plumage rather than the grey, barred pattern on the underside of an adult. It was clearly quite big, with a fairly prominent head but nothing special; long, flat wings, often with the leading-edge curving back smoothly to the tip (but much blunter than a peregrine); rather long tail typically held partly fanned with a rounded tip. It soared and drifted head to wind, going higher and farther off northeast, before turning back and coming closer in a long, accelerating, shallow dive, then a couple of shallow switchbacks with one or two floppy wingbeats, but not developing into real display. Then it soared away again northeast, gaining much greater height – a couple of thousand feet or more? – eventually sparring with a buzzard (which was markedly bigger), before drifting off right out of sight to the northwest.

May. Mostly bright sun with scattered cumulus in blue sky, and the cold wind mostly gone. Mid-morning proved an exciting time (mid-afternoon much less so).

A large, flat-(droop-)winged, long-tailed bird of prey soared up, momentarily suggesting honey-buzzard – but it very quickly became clear that this was a goshawk. It circled upwards, and a distant 'dot' behind it suddenly rose and clapped its wings vigorously over its back – a real honey-buzzard! They moved apart a little, then drifted together and genuinely close rather than just in the line of sight as before. The goshawk gently chased the honey-buzzard, which moved off, then turned, came back and equally gently chased the goshawk. It was minor sparring, nothing of real aggression. The goshawk flew west, the honey-buzzard east, and then moved off in a long, low, steady flight with no more displaying.

The goshawk was remarkably similar at times in shape, and nearly equal in size (or at least, length – wingspan and overall bulk a bit less). It looked very dark – with obvious long head, long, rather narrow-tipped wings and a long, broad (but nearly closed) rounded tail. It sometimes looked very brown above but just occasionally flashed 'pale' below – presumably immature?

It drifted west and rose up again, before being acrobatically chased by a (tiny-looking) hobby! The goshawk soon returned and soared up to a very considerable height, drifted off, then suddenly closed its wings, tilted head-down, stuck out a leg (both?) with its toes bunched like a clenched fist, and plunged down in a stunning headlong dive out of sight into the wood. Meanwhile a peregrine appeared low down: what a place!

Amazingly, just minutes later the goshawk reappeared, soared up again to a great height (giving very long and excellent views although no real pattern could be made out on the dark shape) – then it lifted its tail as if to release the air beneath and reduce lift, pulled in its wings (looking like a giant parachuting meadow pipit), then tilted over, stuck out a leg again and repeated the dive.

Minutes later it did it again. A third dive into a slightly different part of the wood, with one outstretched leg and clenched toes. Then up it came again (amazing!) before it drifted away and this time went into another but more slanting dive. A remarkably obliging bird, giving wonderful prolonged flight views, yet they can be so difficult.

Buzzards sometimes soared up to really enormous heights, just on the edge of naked-eye vision (in perfect conditions), right overhead.

In a variety of places, goshawks have been the main aim of days out, usually in early spring, although I have seen them through the year and even found the odd nest (something akin to an armchair-sized pile of sticks and greenery). They have struck me as really big and obvious or rather smaller and less instantly easy to tell. Once, recently, from my window at home, I watched a goshawk soaring together with a sparrowhawk, an unusual and informative comparison.

Often, they look particularly elegant, belying the oft-repeated description that they look like big sparrowhawks and have long tails and short wings for manoeuvrability inside a wood. Frequently, to me, they look quite long-winged, but it depends what they are doing, whether the wings are at full stretch or relaxed, or drawn back. A recent feature on their identification stressed their longer wings compared with a sparrowhawk, and illustrated them with photographs that demonstrated quite the opposite! In a soar they may look very flat or with the wings slightly uptilted; the tail is moderately long but shorter, rounder and broader than a sparrowhawk's. Anyway, they are good to watch and each time there seems to be something to learn about what they really look like, what they are up to. In a breeding area you may hear them even if you don't see one, with loud, harsh but rather ringing notes answered by thin, whining calls. The quick repetition has variously struck me as like a green woodpecker but less 'squeaky', or a deeper, more even little grebe, but it is quite unlike either really.

Honey and hawks

A New Forest treat: this was a superb hour or so at the end of an early summer morning, with sun through broken high, hazy cloud and little wind. There were, perversely, moments of confusion and moments of brilliance, with some of the best views I've ever had anywhere of honey-buzzard and goshawk:

> I first saw a honey-buzzard low down, almost immediately performing a couple of short bouts of vertical 'wing clapping' displays before going north a little, then swinging up and back west. It was joined by another bird and the two went up very high, practically overhead. The obvious honey-buzzard became less obvious up above, when the wing set etc. was less clear, and with a fanned tail the tail length was not so obvious; I thought these were two honeys. However, one went off west, the other back east – this one was clearly in fact a common buzzard. Then the obvious honey-buzzard reappeared and I had magnificent telescope views. It went up high again and joined three other birds, two of them clearly common buzzards, all of which twisted, turned and dived together although the honey-buzzard was more often just a little separate. It was easy to pick out on shape and because of its largely white underparts – more difficult, though, was a darker bird, looking a little smaller than the common buzzards. But the shape was distinctive enough: another honey. I saw the pale bird again several times, with more excellent telescope views – side on it was very obvious, long-tailed and long- and slender-headed. It went down in a long sloping descent with its wingtips slightly angled back and long tail raised – cuckoo-like in shape (saddle-backed) even without the clichéd head shape.
>
> Incidentally it reinforced my view that the bird seen from our garden last week was a honey-buzzard, too, especially as the direct flight was quite quick, with smooth, quite deep, elastic but quick beats. Usually it soared with the tail closed; the wings could show a slight S shape on the trailing

> edge, or more of an angled leading edge and straight trailing edge effect – typical honey-buzzard shapes, varying according to angle of view and so on.
>
> This paler honey-buzzard now was greyish above, mostly whitish beneath – with a rather dull 'booted eagle' sort of pattern, with dark flight feathers and whitish coverts, dark hood and white underbody, but the carpal area was darker and the greater coverts (I think) barred to give a more blurred pattern. The diagnostic triple tail bars were not so easy from underneath, against the light, but very much clearer and easy to see on the upperside, blackish against pale grey (with a whiter rump?) – the two basal bars very striking. It had a couple of short or missing tail feathers.
>
> At one time I saw both the honey-buzzard and the goshawk in view together in the scope, to the east or northeast – the goshawk came quite close, soared up high, and looked like a big female, quite long-tailed, with long, rather narrow, shapely wings – a very elegant bird altogether – going away, this goshawk and the honey-buzzard could look remarkably alike at times, but the goshawk's longer tail was generally very evident.

A bit of context is necessary if you are in the early learning stage... buzzards are widespread, all-year-round residents, in a wide variety of habitats. Honey-buzzards are rare, summer-visiting birds, mostly in extensive areas of woodland, or in a mix of parkland and scattered forest. In recent years they have increased a little but there are still only around 50 pairs in the whole of the UK, plus just a very few spring and autumn passage migrants heading to and from Scandinavia. Buzzards are easy to see, honeys are difficult: and not always easy to identify for sure. So seeing a honey-buzzard is, or should be, always a real event, a special treat – undoubtedly a GOOD bird by any standards.

Honey-buzzards in June

> A fourth day of brilliant sun and excessive heat, but more cloud. The first try for a honey-buzzard, and near instant success, in the area where I saw a displaying bird last year, but this time it came almost vertically overhead (but very high) and then made off back north. It was immediately obvious – the set of the wings, the slim tail, perhaps the slimness of the body (without the chestiness of a buzzard?) – although distant; but it soon began to soar on flat wings, showing the straight trailing edge and forewing curved back towards the tip, and then gained height and removed all possible doubt (not that there was any) by displaying repeatedly – rising steeply and then

> waving its wings almost vertically above its back several times. It did this while covering a horseshoe course, before going even higher and drifting back over me. The upperside looked dark, the underside whitish on the breast and belly; closer, it showed a broad blackish rim to the wings from the primary tips around the trailing edge to the body; pale primary bases; darker coverts including the primary coverts but without much of a carpal patch; tail details not seen.

On another day I briefly saw a buzzard, then across the road to the south there was a honey-buzzard in the air with a buzzard alongside – success within about two minutes! The proximity of the two was probably coincidence, as they soon drifted apart with no interaction – the honey-buzzard soared around, gained a bit of height, drifted back towards me, then away again, then came back more determinedly northeast before returning south. It didn't display this time but gave excellent views, if never really close. Whether it was the same bird as before I can't say: I noticed each wing on the upper primaries had a prominent pale rounded patch, which I didn't see before, so it was probably a female. Otherwise it was a classic honey-buzzard with a slim body (the rear body lacks the weight and breadth of a buzzard, rather than the foreparts), a thin, protruding head and a longish slender tail, held half open when soaring. In direct flight it had supple, relaxed wingbeats, but at one point made a lot of ground with faster, stronger (but still even and supple) beats. When soaring, the wings were flat with a faint outer droop, or held just fractionally above horizontal. In active flight the leading edge had a fairly marked carpal angle and the leading primaries were angled back a little, while the trailing edge was quite straight; in soaring, the wing shape was more buzzard-like, but it lacked the weight and solidity of a buzzard. Excellent.

LOCAL PATCHES

Chasewater

My local patch for many years in Staffordshire was a large, crystal-clear, canal-feeder reservoir with a canal basin, which I first got to know when we went fishing for beautiful perch and roach. Universally known locally as Norton Pool, or simply 'the pool', its hinterland was an untidy, rough, overgrown, lovely, ugly place full of little dips and hollows, and scarred by the spoil heaps and discarded railway sidings of recent coal-mining activity. As well as being a bit of an adventure playground, it was all brilliant for wildlife of every kind, and I watched the butterflies and identified the untamed tangles of flowers as much as I did the birds, for years. My notebooks were varied, and various field guides were marked up with ticks and asterisks and locations of first encounters. I learned so much. So much has since been tidied up and 'improved' that much of the wildlife interest has been lost, yet other things have appeared 'since my day'.

Had I been around long before, pre-coal-mining days, I would no doubt have been exercised and angry at the despoliation of what would then have been a more pristine patch of heath and marsh. The coal mining dwindled away in my schooldays, leaving so many remnants, but the very roughness added such variety and so many tiny bits of habitat to exploit: the banks, the holes and hollows, the ponds and sphagnum bogs, bushes, hedges, willow thickets... So very much better than corporate tidiness with nowhere for wildlife to go.

Apparently there is now a large sweep of short grass, replacing scrambles of bushes and bramble brakes where I used to watch grasshopper warblers, on which thousands of cowslips have become established. Each spring, at cowslip time, the local authority mows it and destroys the lot. Of course, they know best. They will not listen to the local conservation group – amateurs.

It was Chasewater that gave me so very much early experience. I went on foot, by bike, later by car on the potholed approach. Sometimes I locked my bike to a fence and returned to find a slashed tyre. Once two older men threatened to smash my gears with their catapults. Winter days have a curiously strong place, a powerful presence, in the memory: the best can be so smooth, so quiet and serene. On my bike in December 1964 or January 1965, as a small schoolboy, I would be pushing on through the snow, riding all the way round, looking for birds, while singing *Eight Days a Week* (from the Beatles' new album) silently in my head.

I still sing *Eight Days a Week*, still play the album quite often. It still instantly takes me back to a cold day in Staffordshire, scraping the snow off my shoes, lifting steamed-up binoculars from inside my flimsy jacket, rubbing my red nose with fingers I could no longer feel. A goldeneye, a couple of great crested grebes, a flock of coots, a few reed buntings and I was happy. Back home to the toasted cheese. (Note: toasted cheese, on a plate, not cheese on toast. Just saying.)

It was particularly disheartening one summer, after enjoying a mixed colony of several hundred marsh and spotted orchids, to find a week later that a lorry-load of shale had been tipped on top of the lot. Even in my mid-teens I wrote letters to the local papers and tried to work up an interest in conserving the best bits, but people and conservation organisations had long since 'given up' on a site that had become a focus for water sports and generally less amenable activities. Later, more practical people than I ever was, especially Graham Evans, worked much harder to create a conservation group.

In August there would be migrants in the thickets, as well as wagtails, pipits, wheatears and buntings on the reservoir shore. There would be willow warblers, chiffchaffs, whitethroats and lesser whitethroats, garden warblers, the odd spotted flycatcher. A typical day:

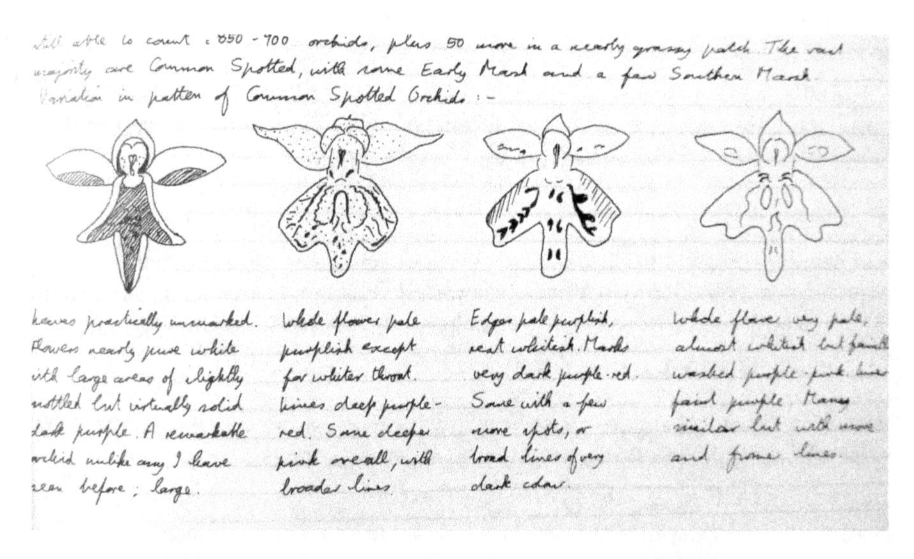

Common spotted orchids, showing some of many variations in flower patterns in one 'colony'.

> The bushes beside the stream at the head of the bay had several blue tits, willow warblers and two or three lesser whitethroats, and I heard the *tchik* of a great spotted woodpecker as well: earlier it had been on a telegraph pole close by.
>
> The bushes east of the dam were quieter than usual but had one or two whitethroats, a lesser whitethroat and a whinchat; nearby areas of rose-bay had two more whinchats and a fine, smart pair of whitethroats. On wires beside the old railway was a wheatear. I think these migrant passerines, although none are rare, are very interesting – and the two whitethroat species in particular, in fresh autumn plumage, look very neat and smart.

They were interesting for sure: far from the coast, just a relative handful of small birds could give the effect of movement and activity in early autumn just as well as at a coastal headland, with birds changing day by day and always the chance of something extra coming along – maybe a redstart! If you have a place like this near you, do make sure you visit as often as you can, to get a sense of the seasonal rhythms that run through the bird world. It could, also, be *yours*... your own little migration watchpoint. And something special, say a pied flycatcher, could be a great highlight and a real achievement, even if it is common enough elsewhere. A wryneck would be out of this world!

If you have a hilltop or other physical landmark, you might find it is good for watching migrants going by during the day. This is 'visible migration', and watching and recording it has become quite a big deal in recent years, with local 'vis-mig' groups (and rivalries) growing up everywhere. There may be flocks of swallows and martins swirling by, or rippling groups of skylarks (or do I mean groups of rippling skylarks?); little dribs and drabs of pipits. Maybe a wheatear or two on the hill, a whinchat in the bushes. Bounding groups of linnets, loosely connected strings of *chup*-ing chaffinches. A big day: perhaps even a ring ouzel.

Just as a reminder, anything can happen. At the same time that I was watching these little and not uncommon migrants, I happened across a couple of waders on the nearby reservoir shore and one was, remarkably, a least sandpiper from North America, which was Britain's fourteenth record. So you can enjoy whatever presents itself and still, now and then, come across 'the big one', if you keep at it. In the way of regular patch watching, I also noted other things: a male pheasant, for example, almost unknown there in an unsuitable area far from decent farmland. What on earth was it doing there? Not quite a vagrant American, but still wholly unexpected.

Hospital Road: KESTREL, 200+ LAPWINGS Wharf Lane YELLOW WAGTAIL, CORN BUNTING
Chasewater
GREAT CRESTED GREBES 1+2(JS) DABCHICKS 3(JS) MUTE SWANS 16+3(JS) TEAL 2(JS)
*GARGANEYS 2(JS) *WIGEON 1♂(JS) SHOVELERS c.8(JS) POCHARDS 6+(JS)
TUFTED DUCKS 50(JS) KESTRELS 2 MERLIN ? 1 COOTS 117+(JS)
MOORHENS c.15 RINGED PLOVERS 9 LITTLE RINGED PLOVERS 3 DUNLINS 5
COMMON SANDPIPER 1 *RUFFS 2 (♂+♀) *COMMON TERNS c.9 *ARCTIC TERN
GREAT SPOTTED WOODPECKER 1 *SWIFTS 5+ HOUSE MARTINS c.75 SAND MARTINS
GOLDFINCHES 20+ LINNETS c.100 The Garganeys have apparently been present for a fortnight or more. Typical head pattern (one darker than the other); both also had typical underwing and upperwing with dark forewings, not pale grey. Small, very fast-flying falcon seen briefly, Merlin or Hobby – probably the former. Water level has fallen even a little lower than before. Now possible to walk across Fly Bay at a point opposite the Sailing Club enclosure (club, not school, which is higher up; upper part of bay just overgrown mud, a little shallow water). More FIRES north of JS, north of the NE bay etc.
SATURDAY SEPTEMBER 11

Even the pheasant was later trumped by my all-time best find, at the same place, just across the reservoir from the least sandpiper spot. It was a Cory's shearwater, from the Atlantic islands, sitting on a shingle shore in bright sunshine on a warm, calm October day. It was the first and for many years the only inland record in Britain. That really does show that *anything can happen.*

My early notes serve to show what you might find if you go to a decent watery spot. The format unashamedly copied that of other people (notably my cousin John), and has stayed unchanged even into the digital era, but you will no doubt find your own best method. My notes include bits and pieces of explanatory commentary, even references to the water level and other relevant details, although I have to say, rarely the weather, which must have been important.

They are interesting to me to go through so many years later but are admittedly often a bit overwhelming, sometimes impenetrable, and mostly just very difficult to put into context without skimming through many more pages. If you are taking notes like this, or become encouraged to start, then finding a way to do a summary of, for example, the breeding season, or winter wildfowl numbers, or the wader passage, or whatever, would be a good idea. It will help in later years should you wish to compare them.

There are little scraps of comments, too, on butterflies, flowers, the odd water shrew. This was my local patch, and I accumulated notes on well over 2,000 visits before moving away from home, with plenty more in a less regular pattern on return visits later. I felt I knew practically every bush. I still feel I belong there more than anywhere else.

One species was especially connected with Chasewater – it was the only regular place for wintering twites in the West Midlands. They were recorded from 1948, usually 10 or 15, sometimes 30, as many as 50 in the 1957/58 winter. For me, in the 1960s, they became special as I began to find them in increasing numbers for a few years, centred on the old slurry tips – flat black pans full of soft 'slack' washed from the coal – and their surrounds. There was an abundance of plants such as wormwood, fat hen and a variety of thistles, sow-thistles, hawkbits and hawkweeds and other weedy, seedy flowers, also exploited by an abundance of linnets.

My first encounter with twites was in January 1966, a year in which the maximum was just 19. In later years I was seeing surprising numbers, 50, 70, 90 maybe... By 1977/78 there were up to 200. By the early 1980s they had gone, all but the odd straggler. Habitat improvement had done for them. But where did they come from, anyway? A few pairs bred on the North Staffordshire Moors; maybe a few more in the Derbyshire hills? Did their demise coincide with the Chasewater decline? Or did the Chasewater decline reduce the numbers that bred? Or were they entirely unconnected, coming from much farther afield? Yet other things are at work, too. In Norfolk, in the 1990s at least, it was easy enough to see a flock of maybe 30 or 50 and then even 100 more pouring down into a distant saltmarsh. They were lovely ordinary little dull brown birds, softly patterned in buff and black and white, with a winter yellow bill and rich tawny-buff throat. Now, scarily, they are not even annual in Norfolk. Another bird that has gone, almost unnoticed, and why is not clear.

The scruffy variety of wet and dry habitat at Chasewater was ideal for a beginner learning his birds, plants and butterflies. I look, for example, at April and May 1967. The shores had a little passage of pied, white (the continental race of pied) and yellow wagtails, many meadow pipits, a few reed buntings. These shoreline groups were always fascinating, tripping along the short grass and on the pebbles where wavelets deposited fluffy bundles of white foam – and the decaying green weed had its own characteristic scent as it dried to a white crust. White wagtails, always smart and interesting to see, peaked at a dozen. There was still a Bewick's swan, goldeneyes, a migrant shelduck, flocks of fieldfares, the first ringed plover on the move in early April, followed by a migrating knot and common sandpipers, 11 dunlins. Displaying snipe would also sit on overhead wires and call loudly. Willow warblers arrived (20 by 22 April), a couple of wheatears, sand martins and swallows, swift by 25 April, a little ringed plover turning up late. By the end of

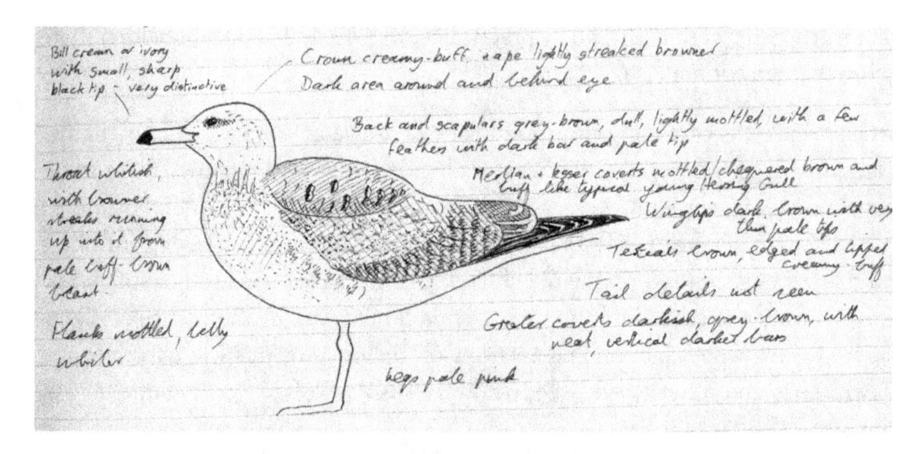

Chasewater was where I first indulged my increasing interest in watching gull roosts. Several strange gulls remain unexplained. I had a vague idea that this might have been an American herring gull... it was not seen in flight.

April, sedge warblers and singing grasshopper warblers. Early May, a turnstone, nine ringed plovers and four dunlins, a cuckoo at last, then 70 swifts on the 4th when turnstones increased to three. A wood warbler, more or less adequately described, but this is one that I have always felt was a bit dubious. Most wood warblers just seem to appear directly at their nesting woods. More wheatears, whitethroats, some passing terns which later in the month resolved themselves into both common and arctic, as well as a black tern; I was yet to work out for myself consistent upperwing differences on common and arctic. On the 9th, a typical Chasewater occurrence, four sanderlings; and there were more later in the month, more turnstones, the first whinchats. I was finding willow warblers' and meadow pipits' nests, as well as nests of little grebe, coot and moorhen.

Over the winter of 2022/23 I heard that my old favourite reservoir was still scoring something entirely new, after all these years – six avocets, swimming out on the water like little white ducks, and extraordinary flocks of 50 little gulls.

Ducking and diving

One November day at Chasewater, when I was still at school and doing my early learning in the bird world, I came across a great northern diver, my first in Staffordshire and the first at that location for 15 years. It had either just arrived by the time I completed my regular circuit, or had stayed hidden near the shore, but suddenly a large bird got up from the water (you don't normally 'flush' divers) and was obviously not going to be a cormorant: very big, powerful, very black and white, rather broad-winged for a diver. The dark flank stripe between the white belly and underwing is always a good clue that it's a diver. It went off far to the northwest and I thought I had lost it, but after it had gone practically out of sight it turned back and came downwind in a very fast and spectacular descent, with its wings arched and angled back, legs drooped, head reared upwards and bill tilted down, creating as near as it could a sort of parachute shape – amazing. It gave excellent views, with the strong pale, bluish bill with a dark ridge, large bulbous head and blackish half-collar all confirming what it was.

Next day I saw it again and took more detailed notes: oddly it was again first seen in flight, but settled despite frequent disturbance by boats. Over two hours or so it gave me some of the best views you could hope for.

On the next two visits I didn't see it but a few days later there it was again: or was it? It was farther away but didn't look just *quite* the same, a bit less sharply black-and-white. The next day, after far better views, it was clear that this was indeed a *second* great northern diver a few days after the first. It tolerated a boat close by but always swam far away from people and was best watched using a car as a hide. It was there again in the next few days but remained extremely wary, staying until 2 December.

On 4 December there was a diver out on the water: immediately it looked different again. Even with my 7x40 binoculars at long range it looked smaller, very dark on top – oh yes, lovely, thank you, a black-throated diver this one! A much scarcer bird inland. There had not been one in the county for 13 years.

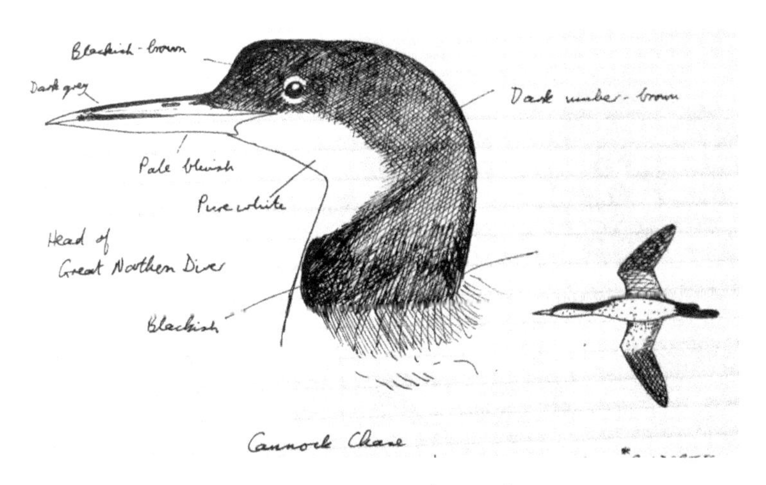

Notes on a Great Northern Diver.

The size difference was especially evident end-on. The two great northerns were big, especially broad-bodied, flat-backed, with an upright head and neck of even width. The black-throat was more delicate, narrower, the back more rounded and the neck slender at the base, broadening at the head. Although large, the head lacked the massive effect of the bigger divers, looking rounder, less elongated, with a more slender, spike-like bill.

A feature of black-throated diver that I think came out later (Lars Jonsson?) is a small greyish upward 'spike' from the dark hindneck into the white where the throat bends into the foreneck. I didn't exactly note it like that, but on this bird wrote that 'the neck looked flat-sided, with a curved "fold" down the side as the bird turned its head to one side, more like fur than feathers'. Over the next few days many pages were full of diver descriptions, even timings of its dives (consistently around 30–35 seconds but in deeper water up to 40). There were short flight views, allowing a further set of comparisons with the great northerns, when it was scared off by a powerboat, but that in the end was that, as it didn't return.

It was a good lesson in maintaining an interest in a 'good bird' that turned out to be three individuals, not one: two great northern divers and a black-throated.

At the same time there were other interesting things around: Iceland gull, of course (always gulls), and two interesting 'scaup' that turned up early in November. At that early stage in my career I had seen few. They were separate from other wildfowl, fast asleep, clearly too large, broad and pale for female

tufted ducks so giving a bit of excitement as they must be scaup: not rare but still scarce inland. But when one woke up, it was a bit disappointing when I noticed it didn't have the right bright white face. However, close and careful examination showed they must be immature (perhaps immature female?) scaup with perfectly round heads (not a hint of a tuft), long, broad bills, even a tell-tale pale crescent on the cheek. The bill on both looked blue-grey with a little black – I wrote, 'on the nail only'.

Tufted ducks have a wide black tip to the bill, which is relatively narrow (but can look remarkably big, especially on a female) while the key feature all the books mentioned was that scaup had black restricted to the nail, the little central 'hook' at the tip of the beak. So, despite the obscure facial pattern, all seemed well. It was more than a week later, after several visits without them, that the two reappeared: they looked exactly as before but the next day I had my doubts. They seemed, after all, to be hybrids.

> The birds I have previously called scaup were present again, with a tufted duck alongside them; in bright sun I had better, longer views than before. The bright light made them look different at certain angles, very often the whitish front patch and pale ear patch were enclosed in an overall diffuse, pale area from above the bill, across the cheeks, under the chin and down the foreneck to a dark band around the lower neck. When facing away from the sun the white face patches practically disappeared; facing the sun, they looked white on one bird, diffuse on the other. The back looked slightly

> more variegated and less clear-cut against the flanks than in dull light. With the face pattern diffuse and pale the impression was much more pochard-like than before; however, the shape of the head and bill was true scaup. They were larger, larger-headed, rounder-headed and larger-billed than a tufted. They had steeper foreheads and rounder, less peaked crowns than a pochard with a definite angle between bill and forehead, not a smooth sweep. The eyes of one were bright yellow, of the other (with the poorest face patches) duller yellowish (trace of pochard??). Both had less black on the bill tip than a tufted, scarcely visible from the side, but nevertheless quite a noticeable triangle head-on. This, plus the pale foreneck, may not be right for scaup. What is scaup x pochard like?

Hmmm... I knew what a scaup should look like, but didn't know what a hybrid might look like (anyway, that was often a cop-out, as hybrids would be highly unusual); nor, at the time, did I know what an immature scaup a few months old would look like. Field guides didn't do that, at the time.

I saw them again in late December (where were they in between?) and they were getting darker on the head with more distinct patches. In January I was at Slimbridge and looked at scaup in the collection, which shed a little light: adult male and female have black on the nail only, but immatures seemed to have a

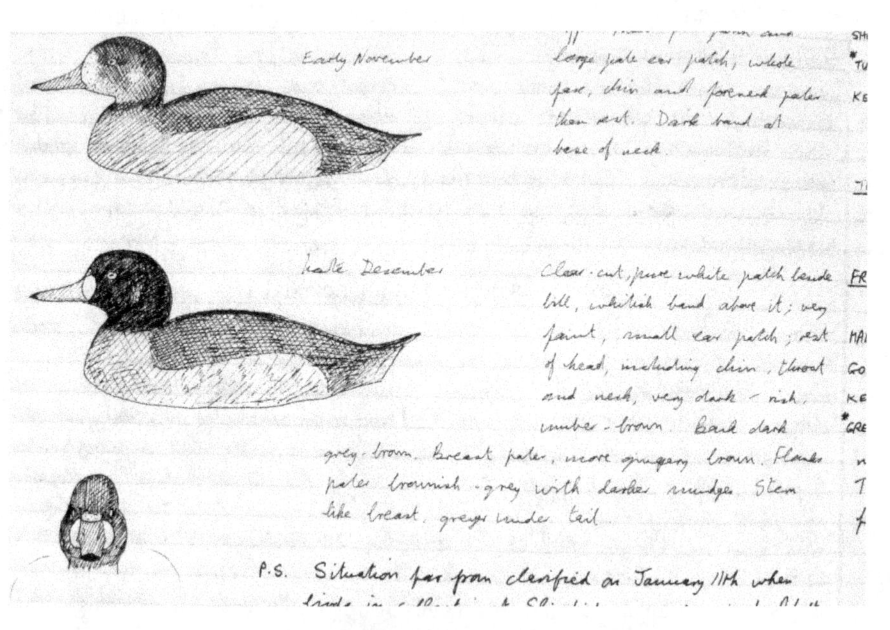

Scaup in early November and late December.

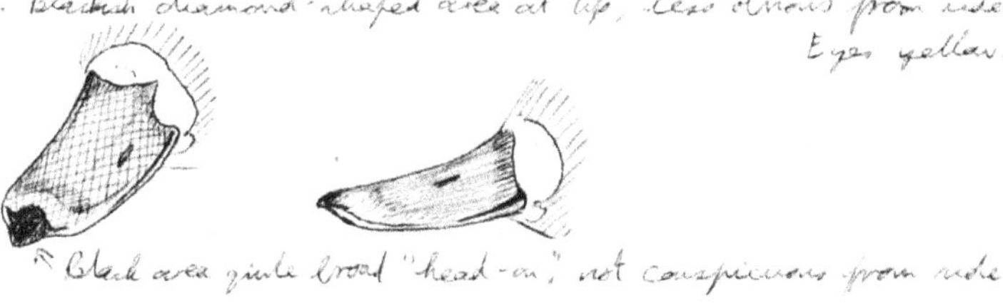

Scaup bills.

dusky dark triangle at the bill tip like these Staffordshire birds. It was early March before I got another view of what I assume was one of the same birds and drew the bill pattern.

But I had obviously been busy. I have no letters now, but noted that Malcolm Ogilvie of the Wildfowl Trust was unable to help and referred me to Dr Jeffery Harrison. A letter from him suggested that there was little doubt that the birds were indeed genuine scaup as he had never yet seen a pale loral patch on a hybrid and the bill characteristics are definitely not so well defined on immature scaup as to be designated one way or the other.

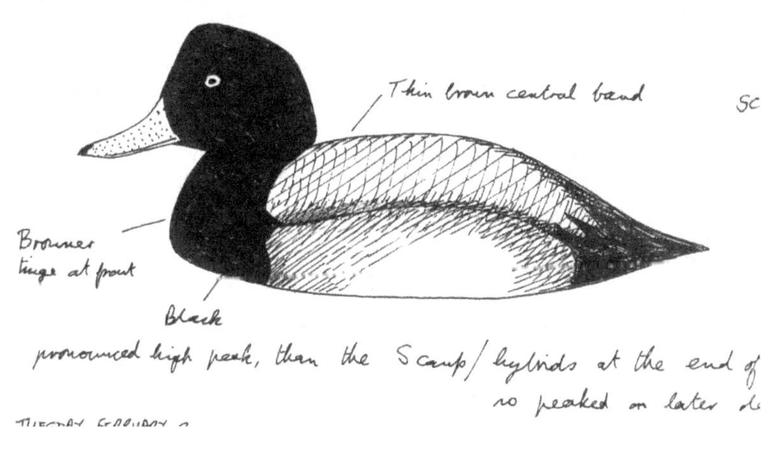

Scaup drake with dusky flanks.

So, a long series of encounters and a variety of notes and sketches later, my scaup had become 'possible hybrids', 'hybrids' and back to real scaup again. It was worth the persistence and another good lesson, even though I had by then had a good deal of experience. Something not so clear from the books I had was elucidated, and now identification guides confirm that scaup have beautifully clear bill patterns except in the immatures, which have dusky triangles.

There was no doubt about another couple of scaup in February, two drakes – both would have been first-winters (a few months old) as they had dusky flanks.

It was many years later that this site produced Europe's first ever lesser scaup, a vagrant from North America. It was also difficult to separate from a possible hybrid at first, but persistence by the finder, John Holian, and my old friend, John Fortey, sorted it out.

Farther afield: Blithfield

Local people would call a reservoir a 'reser-vye'. A French take on 'oir' was not the norm. Chasewater, the canal-feeder reservoir close to me (my local patch) was (despite being more than a mile long) usually simply called by all the locals 'the pool', but my 'second' reservoir was Blithfield, definitely a 'reser-vye', a larger, richer one set in farmland, with two shallow arms at the northern end (falling water level leaving a juicy muddy edge, ideal for waders), a causeway and road across the middle (perfect for watching the gull roost with the telescope on a wall). The deeper, squarer southern end finished against the dam. It was high on the list of inland waters for its wildfowl numbers: an *important* place, one to be proud of.

In drier years the level would fall and the two northern arms became mini-estuaries, brilliant for waders. There was an occasional rarity – I saw, for example, pectoral, white-rumped and buff-breasted sandpipers and lesser yellowlegs over the years – but a good range for somewhere so far from the sea would include ruffs, greenshanks, green and occasionally wood sandpipers, a few godwits, spotted redshanks, sometimes hundreds of dunlins with the odd curlew sandpiper or little stint amongst them... and scores of snipe, too.

It was a good long walk when we set out to do the complete perimeter, starting off from the causeway across the middle northwards around the shallow end, and after lunch doing the squarer, deeper southern end. Blithfield was always a good bet for some interesting birds at any time of year. Inflowing streams would have kingfishers, and there was woodland to add another set of common but

Chasewater (75% frozen over)

GREAT CRESTED GREBES 2 MALLARDS 2 TEAL 5 GOLDENEYES 31
POCHARDS 26+, + 35(JS) TUFTED DUCKS 5+228(JS) *SCAUP 2 MUTE SWANS 34 + 1[illegible]
COOTS 335 + 7(JS) *REDSHANKS 6 DUNLINS 40+ KNOT 1
KESTREL 1 *LONG-EARED OWLS 3 GREY WAGTAIL 1 Long-eared Owls flushed by other birdwatchers, brief flight views plus close range view of one perched before it, too, flew (none, fortunately, leaving the usual thicket - but such disturbance is not necessary. Yesterday a coach party from Derbyshire went to look for them!)
Coltsfoot in flower
. Norton Canes: *BRAMBLINGS c 60 TWITES ? 5 REDPOLLS 1-2

Blithfield Reservoir

HERON 1 *GREAT CRESTED GREBES 239 *GOOSANDERS 107 RUDDY DUCKS 30-60
SHELDUCKS 2 CORMORANTS 47+ REDSHANK 1 DUNLINS 28
*GLAUCOUS GULLS 2 (1 ad 1 1st/2nd year) CURLEW 1 KESTREL 1
*BEWICK'S SWANS 3 *TREECREEPERS 3+ *NUTHATCHES 2 LINNETS 50+
Snowdrops, Dog's Mercury in flower. First good views of these three regular Bewick's Swans - they were grazing in a field north of the Uttoxeter - Abbot's Bromley road, a fine sight. Adult Glaucous Gull seen late in poor light; immature a little earlier but in bad light and visibility I think it was different from the Chasewater bird though basically very similar in colour (no darker secondaries noted - gradual paling towards the rear. Duskier face? More black on bill tip?) Frequently rose in water and flapped; often flew about, usually chased by other gulls, with deep, markedly quick wingbeats despite large size. Bill large and long.

There were three South Staffordshire reservoirs, and I regularly visited two (I never did become a regular at the third, Belvide beside the A5). Here are notes from an icy day at Chasewater and Blithfield. Notice the number of great crested grebes on the larger reservoir: 239! The fact that Chasewater was noted as 75% frozen gives a clue: grebes gather on larger waters when smaller areas freeze, as they must swim in open water and dive for fish. The water may not freeze because of bigger waves, or perhaps busy groups of mute swans and diving coots. If the worst comes to the worst, the grebes have to head off for the coast – and must do so before they are so starved that they don't have enough energy.

interesting species to the wetland specials. All in all, another great place for me to learn my birds.

One early January day was a better than average visit, perhaps.

The usual great crested grebe flock was small, then – sometimes it was well into several scores – but a single Slavonian grebe was an excellent inland find. Cormorants were then rather few; grey herons dotted about (they came here to feed from the colony on writer/TV presenter Phil Drabble's land a few miles away). Canada geese seemed very striking, to me, although friends called them 'farmyard birds'. I recorded just five swans, remarkably, only one of which was a mute: three Bewick's (not altogether unexpected but irregular and a good find) and one whooper swan – which was certainly much more of a surprise, so much scarcer as a one-off winter visitor in the West Midlands. It was also apparently the first time I had seen all three swans in one day, according to my notes at the time.

The 'usual' ducks were in good numbers: at Blithfield, if they were disturbed from the shore or just drifting about ready to feed on the margins or fields later on, then there would be large rafts out in the middle – a couple of thousand mallard, a thousand or more wigeon, a hundred teal, just a handful of shoveler. Five pintails were particularly nice. If they settled on the water they could be watched well enough from the central causeway, with its handy walls at the right height for a telescope.

A dozen ruddy ducks were a sign of what would come in later years, when numbers would regularly go into three figures. Ruddy ducks, native to North America, had 'escaped' from the WWT Slimbridge collection and the Staffordshire reservoirs of Belvide (invisible behind banks but right beside the busy A5), and Blithfield held large non-breeding flocks. Later on, they would be eradicated in a controversial long-lasting and hugely expensive cull, because they had spread to continental Europe and hybridised with rare native white-headed ducks, endangering the future of this already localised and endangered species. Introducing species that have no right to be there is always a risk, and these accidental 'escapes' caused a deal of trouble.

By far the best duck, however, was a single female long-tailed duck, typically a marine bird – except when nesting – that occasionally turns up inland and may stay a while. Usually these wanderers are females or immatures; spanking males with full tails are rare. The more important ones, perhaps, were 50 or more goosanders. Always, given a bit of bright light on a winter afternoon, these would outshine anything else, especially if the water looked dark beneath heavy skies: the males long, low, salmon-pink with black backs and black heads glossed with deep green. Unlike red-breasted mergansers, they have dark eyes (not red) and the bill is deep plum-red (not scarlet); they are exceedingly handsome creatures, and the kind that you needed an interest in birds even to have heard of. Most people would look blank.

There was a good selection of farmland birds in the surrounding fields – finches, buntings, winter thrushes and the like – a couple of stonechats, surprisingly, but also a barn owl. I have some old transparencies showing this farmland landscape

– classic Staffordshire – totally white under a thick covering of frost. Nearby Blithfield Hall is the home of the cute little Bagot goat.

Barn owls were not regulars on my lists, but on the way home I saw two little owls, which were to become quite frequent in my notes. Surprisingly I see none in my local Dorset/Hampshire area now. The barn owl was hunting for many minutes, flying close by in bright sunshine, giving superlative views. Then it perched on a post: fantastic. My notebook says that it would be impossible to get better views.

In the next few days the weather turned cold and there were many birds heading west close to home: flocks of redwings, skylarks and lapwings, joined by 30 golden plovers on one occasion. A local field had 120 yellowhammers with a mixture of reed buntings and tree sparrows, noted without comment: why should there be, when these things were just ordinary, then? They are far from ordinary, now.

***

A different year, an autumn visit: 26 September. A black-necked grebe, obviously very dark-necked and with a peaked crown. Not too common here, a migrant pausing on its way to somewhere else. A goosander, which, unusually, seems to have summered. Only 6 or 10 cormorants (these would become commoner in later years) but more than 30 grey herons trying their luck along the shores. Ducks were generally scattered in small groups, not yet in large winter flocks, but a nice variety, with a few shovelers most interesting. In some years, a little earlier in the autumn, there would be garganeys, a challenge to pick out amongst drifting flocks of teal and gadwalls and showing something of an affinity with the larger shovelers in the way they look.

Waders were good, despite a misty day making it tricky to find and identify them in the farther reaches of the bays that we did not wish to disturb (very occasionally we might have walked out on the mud a little). A ringed plover; a couple of curlews. At least 50 snipe, which belie their usual field-guide-description character and visit areas of wide-open mud on the reservoir shore if the water level is low. Three black-tailed godwits, a good find; a green sandpiper; a spotted redshank, picked up by its frequent abrupt, far-carrying, clearly-enunciated call, *chew it*.

A greenshank, which is more predictable, but a little stint too, another 'good wader' although not uncommon. Only five dunlins, a poor return, but one curlew sandpiper, a really nice bird, although it supplied only a very brief view. Usually the stint and the curlew sandpiper would be there when dunlins were more numerous, but they were always a touch unpredictable – and all the better for it when they did turn up. A ruff... Two black terns, expected but also always good in the autumn ('good bird' is a form of endorsement understood by most birdwatchers, despite being hopelessly vague). Wheatears and yellow wagtails confirmed the autumn date. On some days on the green slope of the dam there would be dozens of wagtails trotting and flitting about, yellow wagtails in mixed plumages with buffy juveniles and females, some males still vividly yellow.

Two great birds, though, made the day. A loud splash. There were big trout here, but that didn't sound right, too heavy. Big brown wings, a flurry of spray – and up comes an osprey! After a lot of flapping and sorting itself out, it flew up and away but luckily curved round, came back, and went by close enough. It was seen again later, mobbed by lapwings. It was as if British birds hadn't quite got used to ospreys yet, hadn't acclimatised to them as they increased. An osprey over a reservoir still usually causes mass panic amongst all other species – yet is the most unlikely bird to pose any danger to any of the others. It was seen again a few times, once hovering, later carrying a big fish. Pale barring on the upperpart feathers – crescentic buff fringes – showed it must be a juvenile, hatched earlier that summer in Scotland or maybe Scandinavia, on its way to Africa. Such detail in my notes reminds me it must have given a good view.

And two frequent colleagues on these reservoir walks, John Ridley and Alan Dean, called out 'great skua' as a big, dark bird appeared far off to the south, on the water: not flying, but swimming, a great spot. It flapped a few times, but was evidently busy pulling apart some sort of prey: apparently a dead gull. But, still hungry, it suddenly flew low for 30–40 yards, aiming straight for a coot: the coot dived for its life, the skua settled, then dashed at another coot nearby. This coot relied on its usual defence, not diving but leaning back and kicking out with its big feet, but against a great skua this was too feeble. It was a fatal decision. The skua hit it, breast-to-breast, using its size and weight, then repeatedly banged at the coot with its bill. End of coot. There had been very few records of great skuas in Staffordshire, and this was a calm day, not during or after gales when odd seabirds might be looked for.

As ever: unless you go out, you won't see anything. If you do, who knows what might turn up? Just once in a while... and the more whiles, the more onces come along.

Blackpill and Gower

For years I had been living on scraps so far as coastal species were concerned. Well, I was still in my teens anyway. Staffordshire is as far inland as you can get. Spending long periods over six years in Swansea changed that completely. The views for a start: across Swansea Bay beyond the docks to the steelworks of Port Talbot, the other way along the curve of the bay to the long, elegant headland of The Mumbles with its little lighthouse and purple sandpipers. Directly south over the wild waters of the Bristol Channel to the high ground of Somerset and Devon. And from the very beginning, flocks of gulls heading for the beach and swirling lines of oystercatchers manoeuvring with the rising tide.

It was not long before the high tide roost in the western half of Swansea Bay became obvious: Blackpill. Here the 'pill', a shallow stream, flowed out from Clyne Valley and crossed the beach beside a long, low sandbank that remained uncovered after most of the surrounding beach was awash. If the bar was swamped, too, there were always the playing fields close by, where large flocks of some waders and gulls would gather.

It was not a case of finding a place for myself, any more than Chasewater had been: it was well monitored by local birdwatchers and I simply muscled in a little bit on the side. Although, thinking about it now, it was thoroughly well documented by Bob Howells, who took it upon himself to do the wildfowl and wader counts for Blackpill and the whole of the Burry Inlet for many years, but otherwise was surprisingly little visited by local birdwatchers. I rarely saw anyone else there apart from us students. But for me, Blackpill became a second home – and farther afield, out on Gower and the north Gower marshes of the Burry Inlet, the wild habitats and great birds were just a joy.

Blackpill was chiefly known for its waders and gulls. Now and then something else would crop up and cause a stir, perhaps a snow bunting or two, or a diver offshore, maybe a passing peregrine. The wader roost, though, was always fascinating and often spectacular: hundreds or thousands of oystercatchers and dunlins, hundreds of ringed plovers, sanderlings, relatively few knots, scores of bar-tailed godwits (but very rarely a black-tailed), hundreds of curlews, grey plovers, redshanks... There was always the chance of something better, but only very occasionally would there be a rare wader, such as a Kentish plover.

Gower had its small seabird colony on the mainland shore at Mewslade, plus a larger one on the long, narrow island of Worm's Head. Offshore were flocks of common scoters. At the outer end of the Burry Inlet estuary, opposite Burry Port, were special birds such as brent geese and eiders, as well as grebes and divers coming in and out with the tide. The dunes and marshes were hunted by peregrines, hen harriers and merlins. The marshes had great flocks of wigeon and pintails, flocks of golden plovers; the mudflats hosted many thousands of oystercatchers. Bob Howells remarkably counted them all. Gower woods had a great assortment of small woodland species, while the reedbeds of Oxwich were full of reed warblers and sedge warblers, and attracted the odd Cetti's warbler early in the colonisation of Wales. We saw an occasional bittern, even a purple heron, and there was sometimes a wintering great grey shrike.

All in all, this was a sensational area for anyone interested in birds, let alone its fascinating array of wild flowers and butterflies, diversified by the limestone of Gower's cliffs. There was always something to see.

Blackpill's beach usually had a few thousand black-headed, herring and common gulls, some lesser and great black-backs, and the odd yellow-legged gull from time to time before anyone even called them that. Kittiwakes gathered in large flocks in summer, miles from the nearest colony, with a number of two-year-olds and faded one-year-olds showing a great variety of interesting plumage stages. Then there were a few Mediterranean gulls most of the time, giving me endless opportunities to scribble notes. I've drawn scores and scores of Mediterranean gull heads! Little gulls were scarce but not altogether unexpected. Glaucous was rare, Iceland a bit rarer still, but unusually both turned up occasionally in summer. And ring-billed gulls from America appeared in a series, several birds over a few years, comprising the first few British records. It was a gull-watcher's paradise, to be sure, and I was lucky enough to share this for a short time with other enthusiasts such as Keith Vinicombe.

The first ring-billed gull was an adult that I picked out in March 1973; Keith found the second, a very difficult first-year, later that summer. On 5 December Keith pointed out a pale gull on the beach that looked 'right' for a ring-billed but seemed to have a dark eye, which was clearly wrong. On 15th the same bird was there again. I was there with Derek Thomas, a maths lecturer at the university.

> Before reaching the beach I noticed an 'odd' gull and pointed it out to Derek – this indicated how distinct it was, clearly 'different'. We had excellent views from 75 yards for about 15 minutes. The light was dull, even, the bird standing side-on with a few adult common gulls nearby.

So followed pages of notes... a perfect adult ring-billed, except –

> The eye was the only problematical feature. At that range, eyes of herring gulls were seen to be yellow only with some difficulty, although clearly not

> dark. Eyes of common gulls looked black. This gull had an eye that at first looked very dark, but the more I looked with my telescope the more obvious it became that it was less solidly black than that of a common gull. In fact, it showed a dark ring with a clear black pupil in a paler iris. The iris colour could not be made out but it did not look very pale.
>
> A second-year ring-billed will have a dark iris: at what point does it become pale?

This ring-billed was seen rather infrequently through the winter, especially well in late February and again in mid-March. It stayed into April but then, on 3 April, there was a second-year ring-billed. So, at Blackpill there had been an adult 14–31 March 1973, a first-year 3–14 June 1973, an adult 5 December to April 1974, and a second-year 3 April 1974. The second-year could, of course, have been the earlier first-year, now a bit older.

There was a gap from 3 April to 27 April 1974. My notes and sketches proved useful, as on 4 May I saw this bird again and clarified what I had already been thinking: this was another one. It seemed to be a second-summer individual, somewhat more advanced than the early-April one, with detailed differences in the wing and tail patterns as well as the head and bill. This was just an exciting time for gulls in Swansea Bay. A similar sequence would be repeated many years

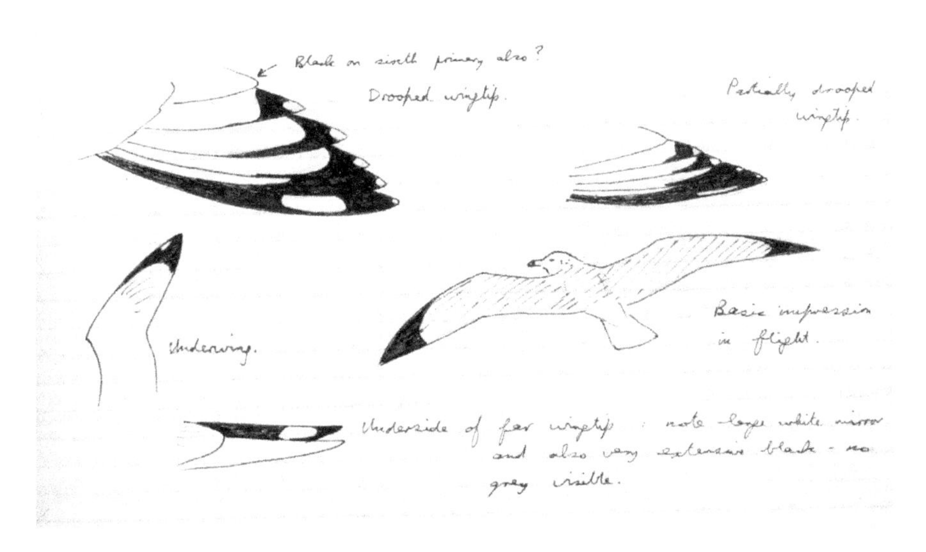

Details of a ring-billed gull. At that time (in 1973), it was not in our European bird guides.

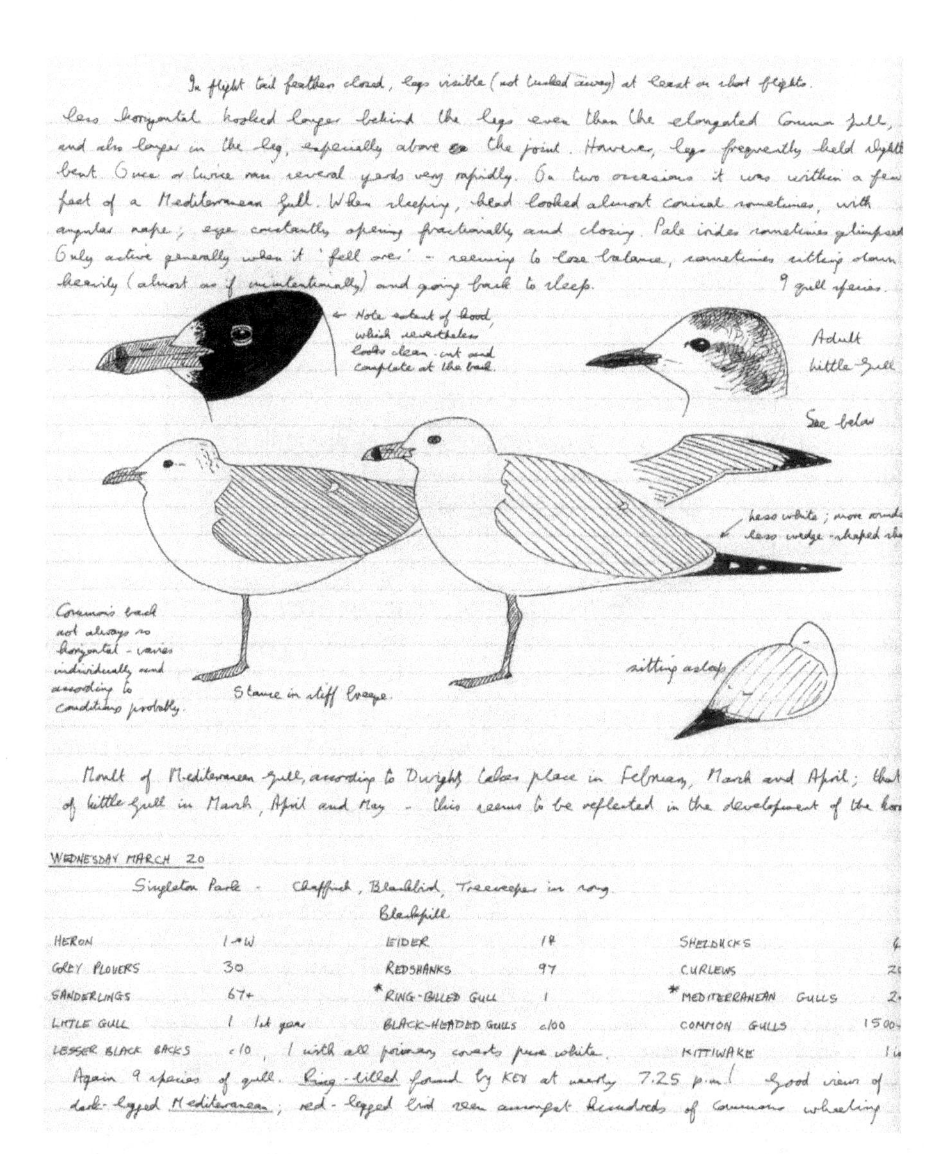

Gulls on the beach near Swansea: Mediterranean, little common and ring-billed (from America).

later at Blashford Lakes in Hampshire, but these early birds were just so very special. It was a while before the species was seen anywhere else in Britain but by 1981 there were real influxes becoming apparent. Remarkably I bumped into one in Argyll, in a roost at Lochgilphead, after a visit to Islay just a few years later (March 1984).

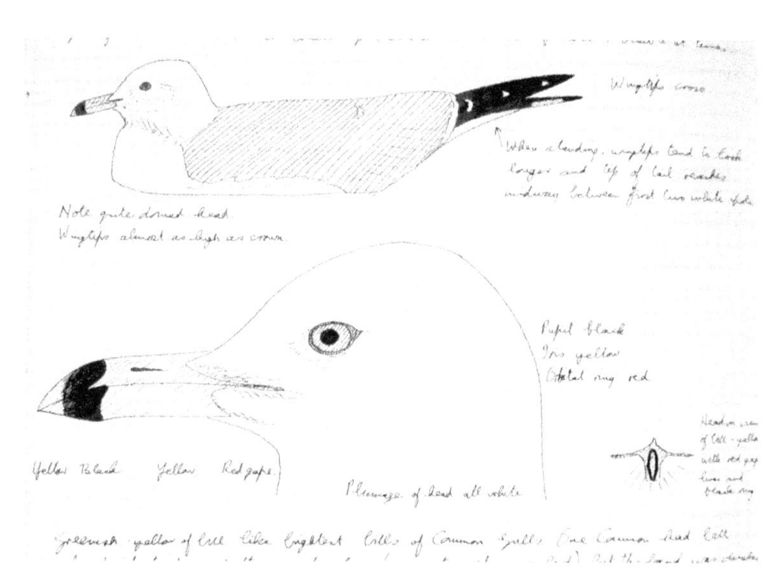

Ring-billed gull seen unusually well, for once not miles away on a beach, or in a roost as it was quickly getting dark.

More on ring-billed gulls... a second-year bird in South Wales.

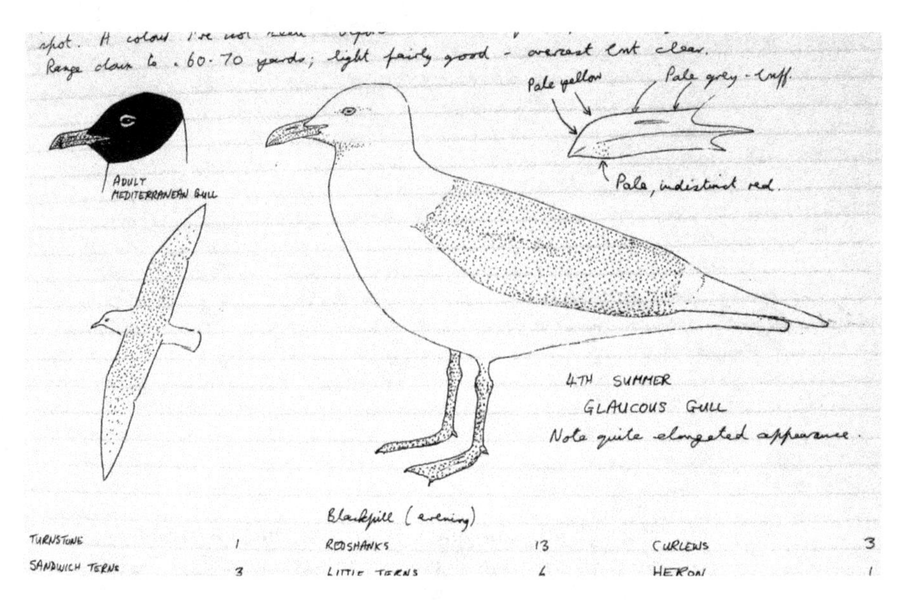

Another odd combination: a breeding-plumage Mediterranean gull and a glaucous gull.

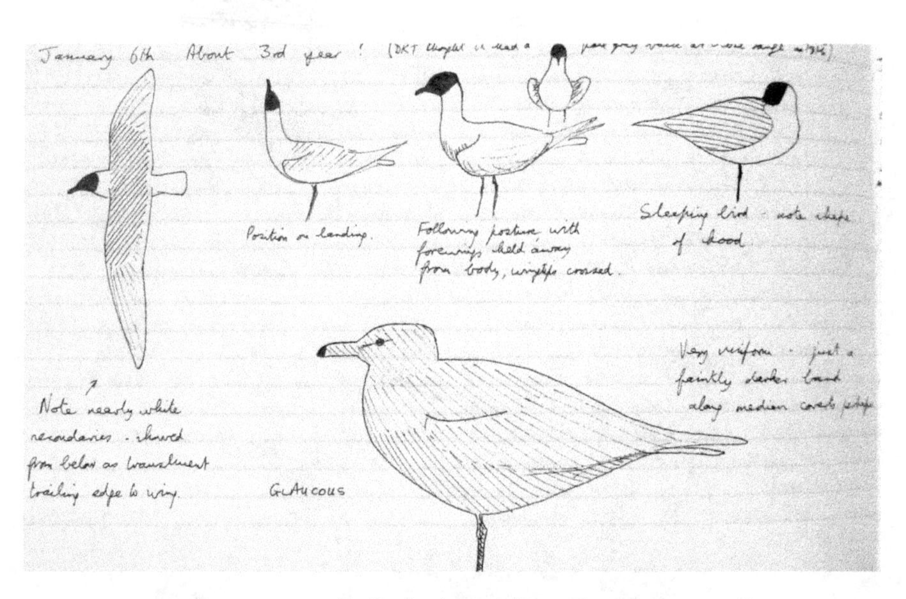

Mediterranean gulls displaying, with another glaucous gull close by, all on the Swansea Bay beach, early 1970s.

More ring-billed gulls: note the bit about the common gull's eye. I had probably written 'common gull, eye black' but it is actually dark brown, while a ring-billed develops a clear, pale yellow iris.

Out on the estuary

If you can get to a coastline, you will usually see so many more birds. My few years near Gower, visiting the fabulous Burry Inlet on its northern shore, and sampling the joys of Swansea Bay, added so much to my bird experience and embryonic knowledge: so many wonderful days.

Early December. Instead of wandering around at the western, outer end of the estuary as we usually did, with Peter Garvey I explored part of the upper estuary. It was in the news at the time as the local cockle-picking interests were up in arms against the oystercatchers and wanting a cull. The estuary was indeed a focus for oystercatchers, those big, noisy, excitable, carrot-billed black-and-white waders that add so much to the estuarine experience. There were suggestions of a cull of 10,000 birds that would cost about three times the annual cockle revenue of £35,000 at the time.

Indeed, on this day we soon came across oystercatchers, as you would expect, and my notes record 10,000+. The next line also had 10,000+, an unusual double-entry for my records: this was a flock of knots. Dunlins were next with up to 2,000, curlews 500. After that, 330+ was the biggest number, against *pintails; just beating 300 linnets. More unexpected were 14 white-fronted geese, while wigeon reached a mere 10 or so (they would have been farther down the estuary somewhere, in their hundreds).

Most unusual was a whimbrel, one of several of my records rejected by the local bird club, but it called loudly with its usual rapidly repeated note: actually quite a good, interesting record for December, had it ever made it into print. Rejected records were sometimes an annoyance. My first roseate terns, also seen by other people (a group of Gower birdwatchers happened to be there for once), an adult still pink-breasted in early October, and a juvenile dark-crowned with a black bill and dark brown legs, were chucked out with no explanation, while a local bird report included an accepted one with a description that proved it was an arctic tern. Another published, accepted report not far away was of a male pied flycatcher catching insects as it jumped from stone to stone in a stream, probably wagging its tail, in mid-winter.

There was a good scatter of lively finches – greenfinches, bullfinches, linnets, goldfinches, chaffinches – and buntings. Sunshine picked up the fierce yellow of male yellowhammers, especially appreciated. It was just a good afternoon on these flat marshes and the adjacent slopes. The marsh is intricately etched by little, steep-sided muddy channels, which would lead you round an apparent easy route on the vivid emerald turf only to reach a dead-end against a deep gully, so a short walk would become a very long one with innumerable setbacks and retracing of steps: often, the only way out was the way in. It was never worth the risk of being caught by the tide.

I think the pintails were then the biggest flock I had seen: pintails are widespread in smaller numbers but a few bigger estuaries have substantial flocks. It was, however, the waders that left the biggest impression. Stream after stream after stream of oystercatchers poured out towards the outer estuary, as the tide fell. It was the knots, though, that proved impossible to count properly: a continuous shoulder-to-shoulder mass on the ground, or a twisting, spiralling cloud in the air. As they weaved their endless patterns, they showed the typical textbook flashes of white and sudden change to soft grey, momentarily almost disappearing, then deep, dark shadow for a few seconds before the next twist and shine. The 'cloud of smoke' metaphor was appropriate, but I felt 'a passing heavy shower of rain' was more accurate, especially with the naked eye. Either way, I said the flock defied description, the finest bird flock I had ever seen. At that time, I had not seen even bigger wader flocks elsewhere, nor gigantic million-plus flocks of starlings.

At the other end of the estuary is Whiteford Point (say 'Witferd'), a long arm of dunes holding back the marsh with a mussel scarp at the end marked by an ancient cast-iron lighthouse, its high balcony providing a roost for quirkily handsome, posing cormorants. From the bus stop at Cwm Ivy we walked down a sloping path to the base of Whiteford proper and out through a dark pinewood, past the little offices of the Nature Conservancy. On a cold, wet or windy day, the wood provided a little quiet shelter, a lull before the storm. On one side were fields with ponies and a patch of rough, wet snipey (occasionally jack snipey) grass, then saltmarsh to the right with its network of watery creeks and dunes to the left, with a long sandy beach overlooking Carmarthen Bay and out past the rock of Burry Holm to Worm's Head. (Jack snipe was one species that we students would occasionally report that Bob Howells did not. One day a couple of us were in the long grass, clapping hands, when Bob appeared along the path. 'Ah, this is how you do it, is it?' he said – putting us in our places, probably quite rightly. But you never see jack snipe otherwise, and they only go 20 yards...)

The marsh continues all along the north Gower coast through Penclawdd and Llanrhidian, to end against this right-angle of sand, and the estuary fills and empties via narrow channels around the tip of the dunes. Looking into the marsh and along the estuary, we could enjoy the regular waders, mostly at long range, but with other things occasionally popping up in the creeks, such as spotted redshanks. But on the swirling, fast-running tide, there were many birds to see if you could only stop your eyes watering in the bitter wind.

Whiteford was famously the only place for hundreds of miles that had a flock of brent geese each winter. They were remarkably elusive at times: never many, perhaps 20 or 30 on an ordinary day. It was also a regular spot for eiders: one day, for example, I counted 24 males, 29 females, and it was here that I learned a good deal about them. We watched for red-breasted mergansers, occasional common scoters, sometimes even a velvet scoter. There were usually black-necked grebes, outnumbering one or two Slavonians (sometimes more), rarely a red-necked. Red-necked is practically a rarity now, UK-wide. Sometimes the grebes were remarkably close, bouncing on the choppy sea, brightly cherry-eyed. Maybe a diver: great northern, probably. Now and then a long-tailed duck. Over the marsh, with a bit of luck, a hen harrier, or a short-eared owl; sometimes a merlin, or a peregrine. The harrier would be most likely.

It could be a long hard slog walking out and around Whiteford, tackling stickily muddy paths at first before the sand; sometimes deep in snow, sometimes freezing cold but brilliant in winter sun. You could stand on the undulating wet sand on the beach near the lighthouse and make the whole thing shake alarmingly underfoot. At the end of the day, there was the climb back uphill to

Cwm Ivy, where I would turn round and look back out across the marsh, and over the estuary to the giant chimneys of Burry Port and distant Llanelli.

This was an irreplaceable moment: it could be so quiet, so still, so extraordinarily tranquil, even with long, rolling waves washing against the lighthouse and along the western beach, with their distant, half-heard, half-felt rumble. Like birds, at times, these things can be especially good at a distance, wild, empty and remote, impossibly serene. Tranquil and serene are always the words that come to mind. I don't know if it was a bit odd, or unusual, to be so affected at the end of my teens; I was, at other times, also quite capable of turning up the volume on the Rolling Stones, Pink Floyd or Bob Dylan.

Those winter days on the Gower marshes remain such precious memories. In later years I had great times in summer, with the flowers and butterflies too,

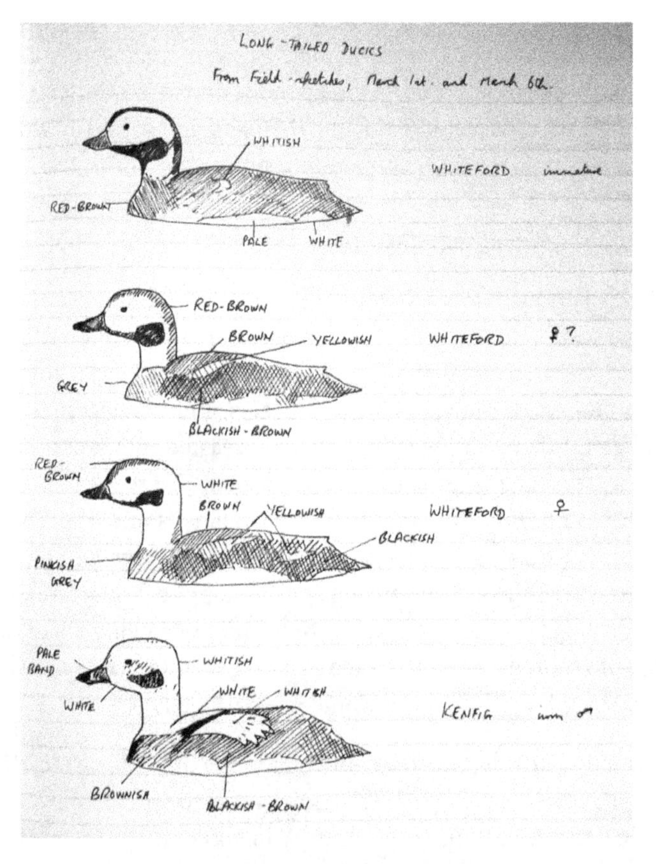

Here are notes on long-tailed ducks at Whiteford Point and at Kenfig Pool, a few miles east of Swansea.

a lizard posing for a photograph on my foot; but it was never quite the same, on my own. Dune slacks were full of colourful marsh and spotted orchids, less eye-catching marsh helleborines, and they offered the chance of a difficult, tiny, rare, 'green' fen orchid. With Peter Garvey, I would return on the bus, dash down through the park and, according to my diary (but now forgotten), cut sneakily through the vicar's garden and over his fence, to get back to the snack bar in time for the last pasty and beans. If we missed it, it might be potato croquette and chips. History (my diary) does not record whether or not we were ever accosted by said vicar. One autumn, long after I had left Swansea, I returned with a birdwatching group out there and came across a vagrant Richard's pipit, a barely believable moment.

Looking at the recorded details, now, without over-analysing things, it is clear that many things are involved. From large to small: nature, outdoors, Wales, my few old, long-lost university friends, uncertain connections with the local bird society whose members I now admire so much more respectfully, simply learning my birds... an affinity, still, for many things Welsh. Dylan Thomas, Philp Madoc, Max Boyce, Anthony Hopkins – all those great voices. Lost youth, too, for certain.

Thinking back about birds can do this for you. But mainly they were just good days with good birds in exceptional situations, unspoiled, undeveloped, often near-pristine... and unrepeatable. Nostalgia sums it all up, true, but it remains in my mind as something very special: a time, a place, good companions, a period of my life now long gone, and great birds – that will always be great.

Ravensong

> Swansea. 29 October. Two ravens flew over the park, then four, behaving as two distinct pairs, settling high on isolated tall pines. A good deal of calling: usual croaks and rasping noises, then a throaty 'hoot' with a kind of 'voice-breaking' catch in the middle, changing to a higher note. Two tapped their bills, raised their wings, fanned their tails while still perched and gave a great variety of weird noises: rasping and retching, rattles, with a short, hollow rattle like the drumming of a woodpecker, extremely similar in quality to a great spotted's drum; some honking and crooning, somewhere between a pigeon and a turkey's gobble...

Ravens also gave a long, rambling 'subsong' in spring, barely audible except at very close range: I listened from just below their tree perch. These may well have been the pair that nested for a time on the square, grey tower of the Guildhall in

Swansea, unnoticed by ordinary passers-by or the outflow of distinguished local politicians, or the resonant male-voice choirs in their tuxedos before a show, but always in full view on their massive stick nest.

Little Paxton

When I moved to Bedfordshire in early 1978, I found a new 'local patch' in the shape of Little Paxton pits on the edge of Cambridgeshire. Just up the road was a much larger reservoir, Grafham Water, but for some reason I was less attached to Grafham and it became a sort of addendum to good days at Paxton even though, in the end, I did see a good many more unusual birds at Grafham over the years. Between the two, though, they made a kind of substitute for the Staffordshire

THURSDAY MAY 3
Little Paxton
*GREAT CRESTED GREBES SN – 1 pair on the river HERON 1 – pless young heard in nest?
TUFTED DUCKS c.50 GREY LAG GEESE 2 RINGED PLOVERS 2-3 (display flights etc.)
*LITTLE RINGED PLOVERS 2-3 *OYSTERCATCHER 1 REDSHANKS c.4 COMMON SANDPIPER 1
RED-LEGGED PARTRIDGES 7-8 BLACK-HEADED GULLS – reduced numbers. One white bird (see below)
COMMON GULL 1 1st yr. *COMMON TERNS 2 *SWALLOWS 20-30 *HOUSE MARTINS 10-20
*SAND MARTINS c.10 *YELLOW WAGTAILS c.10 *WHITE WAGTAIL 1 ♂ LESSER WHITETHROAT 1
WILLOW WARBLERS SN *SEDGE WARBLERS c.5 + several heard NIGHTINGALES 1 or 2 heard
REED BUNTINGS SN YELLOWHAMMERS SN *BRAMBLING 1 ♀

Albinistic Black-headed Gull – size and shape normal as far as I could tell but white plumage made even direct comparisons very difficult. Bill pale orange-red with darker tip and base; eye dark; legs pale orange, even yellow-orange. Entire plumage unmarked white. The head was less 'brilliant white' than the breast but could not be said to anything but white and there was no trace of any dark marks near the eyes or on the ear coverts. Whole wing above and below pure white, but more silvery or 'frosty' in flight.

I spent some minutes watching up to six Swallows perched in a waterside oak, often singing, in bright sunshine. Using a telescope made one bird fill the field of view. There can hardly be anything in the world to beat them!

SATURDAY MAY 12

*Here's an interesting mix at the Little Paxton complex, set in largely agricultural country, just off the A1, alongside the Great Ouse. It has since seen a good deal of housing built on nearby fields. It was, for a start, not normal to see a brambling in May, when listening to nightingales. Swallows would have been unremarkable except that they settled on a tree and I looked so closely at them with a telescope; but 10 spring yellow wagtails would now be very welcome, as they are not so frequent near my Hampshire/Dorset border home. The male white wagtail was of the continental form (subspecies) of our familiar pied wagtail, a typical spring migrant (usually in April) but not ever so common generally. Given a good view of a smart male, it got the * treatment.*

Chasewater/Blithfield duo, although I must admit that many a weekend was spent back in Staffordshire, back at my original haunts. Nevertheless, Paxton was visited sufficiently often for my coverage to count as 'purposeful', with a bit of order and discipline about the notes and counts.

Paxton had a heronry, and cormorants began to join in and breed in the tall trees, too. Now there are also egrets, I believe. Black-headed gulls and common terns nested. It was always a good place for nightingales in a mix of hazel and bramble thickets on gravelly substrate and various willows, alders and other dense shrubs around the lakes. Now and then a nightingale would fly up and sing from an exposed overhead wire.

April/May 1989 gives an idea of what might be expected then, together with some extra goodies:

27 April, Grafham Water

*Surf scoter	1 immature male
Little gull	1
Common terns	70–80
Yellow wagtails	5

The scoter stayed a few days, a remarkable occurrence for this rare North American duck that usually appears on the coast. At the beginning of May I checked the Paxton herons and found a mix of adults sitting, small chicks, large chicks and fledged young. The Grafham superstar scoter was still there. A dash east farther into Cambridgeshire produced a fine pair of stone-curlews but, more unusual for me by far, 23 or 24 dotterels. But back to Paxton: both ringed and little ringed plovers, oystercatchers, dunlins, common sandpipers and a greenshank made a decent wader showing. A female scaup was particularly good for the time of year (the best duck I saw there, a year or two earlier, was a drake ring-necked duck). A little gull was more to be expected, perhaps. But the 'ordinary' stuff included two cuckoos, three turtle doves, sedge and willow warblers and whitethroats, yellow wagtails with one male of the continental 'blue-headed' race (certainly unusual), 100 sand martins, nightingales, a decent influx of swifts. Nice things, and now mostly much reduced. Turtle doves, especially, seem to have gone altogether. There were occasionally still tree sparrows at Paxton, too, but they soon disappeared. It is hard to explain their decline, short of some natural retraction of range in a species that has fluctuated before.

Later in May black terns were on the move through Grafham, a grasshopper warbler sang at Paxton... but the best bird by far was a splendid immature male red-footed falcon. It was, remarkably, the second time a red-foot had been found here on an RSPB Dawn Chorus walk on the May Bank Holiday weekend. It was just 35 yards away in full sun, but didn't look ever so well: it kept gaping, throwing back its head and 'coughing', and was a bit ruffled. Yet I noted that its black eyes were particularly bright and gleaming with a shiny, intensely red orbital ring. It was not quite a full adult male but nor the standard 'first-summer' male that often turns up in the UK. In a third spring I saw a female, too.

At such close range it had a particularly obvious droopy-billed look, almost parrot-like (distinctive on some photos of the species but not always so). Its legs were clearly on the yellow side of orange, not at all red. One of the Paxton red-foots was found by an observer who was colour-blind and needed a bit of help with the leg colour.

The following day, the red-foot was still there but, more surprising again, there were four hobbies in view all morning. This in itself was unusual; to get them together like this was fantastic. All five flew around near the heronry, catching insects in their feet. It was nice to be able to compare their shapes:

> Red-footed very like hobby in shape and actions but a fraction smaller, more lightly built, slightly narrower across the chest, narrower and more 'waisted' at the base of the tail; tail narrower, longer and more wedge-tipped (hobby's square). Wings not quite so sharp or swept back on average, but wing shapes of both very varied. Flight a little less relaxed than hobby, a little less assured, just slightly more flappy and kestrel-like whereas hobby looked more peregrine-like. But both very similar really!

Odd things would appear at Little Paxton, rather than real rarities. For example, that same year on 30 July, not only a black tern and two little gulls, but a juvenile kittiwake. A peculiar date for an inland kittiwake. A few months later, Grafham repeated an old Chasewater trick, with a long-staying great northern diver that one day changed to a black-throated. Paxton, meanwhile, did well for wildfowl: one February day, as a random example, produced 40 shovelers, 800 wigeon, 200 gadwalls, 200 tufted ducks, a scaup, 400 pochards, two red-crested pochards, 60 goldeneyes, five goosanders, two male smews and a couple of the dreaded ruddy ducks (since eradicated from the UK). That was quite a haul: scaup, red-crested pochards, smews, any of these would have been great to see, but all together, that was especially good. A few days later the smews had changed to one male and one female, while scaup increased to four, although they were quite separate on different pits. Like gulls, flocks of ducks always repay good, hard scrutiny, just in case.

The female red-footed falcon.

There is a reminder from August 1994 of how recently and quickly some things have changed.

> Something unexpected at Little Paxton – a rare event. A little egret. It flew off but fortunately returned later and I had good views... I think this is my first inland and, although there are flocks of them on the south coast, little egrets are still very rare up here.

My first UK little egret was long before in South Wales, but I had never seen one in midland England before, and this one still received close scrutiny and a full set of descriptive notes.

On the same day I also noted:

> The two hobbies were wonderful. One was high up at first, catching insects, with periodic steep dives or long, very fast chases ending in skua-like aerobatics. It may have had pale tips to the secondaries but otherwise looked grey above, quite slaty, paler on the rump and tail base [*a feature I later came to repeat in books as characteristic of the peregrine as opposed to a hobby, but it varies a bit*] but its head was dull, the underparts quite buffish, with no red evident. The inner primaries looked a little paler than

> the outer few, and shorter, giving a slight irregularity, as if in moult, but I presume it must have been a juvenile. After a while I became aware of a second, over trees closer to me. It was a superb adult, with red easily visible in flight, although quite pale. It frequently settled on some dead branches, where I enjoyed easily the best views I can remember of an adult hobby, practically as good as a very tame juvenile at Titchwell last autumn. It sometimes dashed out, sometimes going very high but usually keeping low, chasing dragonflies. It put on a wonderful show, mixing elegance with real speed, agility and aerobatics, all with a sense of purpose and menace. Its head pattern was beautifully crisp, with a dark grey forehead and crown above a pale supercilium and a very black moustache. The splashings of black in long, broad lines down the underparts, with scaly white tips to the flank feathers waving in the wind, made a beautiful pattern; the feet were vivid yellow, with intensely black claws.

Like most such places, Little Paxton would sometimes have a little purple patch, with everything turning up at once. Such a spell was around the turn of the year from 1994 to 1995. In early December there had already been a great northern diver, itself a bit of a 'mega' for these pits. On 21 December it was joined by a black-throated diver. In fact, I had not seen the great northern despite trying hard since 3 December, but now it was there again, clearly the same individual, and literally alongside it – within a couple of feet – was this smaller, slimmer bird. Unusually, I was there at dawn, for half an hour in the morning mist before having to go to work, but at the weekend the black-throated was still there. After a few days of mist and mugginess, the day dawned bright and sunny with a complete covering of hoar frost, still covering every bud and twig and blade of grass for the whole day: astounding. The diver gave marvellous views on a half-frozen lake, more or less against the light but still... After Christmas the two were together again, while a drake smew and a red-crested pochard (of uncertain origin but looking good, all the same) had appeared. A dash back to the Midlands also resulted in a smew at Chasewater, where they are far less regular. By early January Paxton was still scoring, with the great northern diver, 1,000 wigeon, 100 gadwalls, 250 pochards, 500 tufted ducks, a scaup, 80+ goldeneyes, 16 goosanders and a smew. A few days later, unusually, mandarins added to the list of wildfowl.

Grafham was particularly good one February day in a different year. For a start, 500 great crested grebes. These grebes are generally forced off smaller bits of water as they freeze and concentrate on the larger lakes, keeping free of ice. With them, there was a single Slavonian grebe. Sawbills totalled 30 goosanders

Black-throated diver, Little Paxton.

and two smews. There was a seemingly obligatory female scaup... but far more unusual, a velvet scoter. And two ruffs, strange for the time of year.

> The great crested grebes included slightly separated loose groups of 30 or 40 near the dam, which were harried at times by common gulls. There were several goosanders with them. The grebes were not diving much, but, because there were so many, every minute or two there would be one popping up with or without a fish. Each time, one or two drake goosanders would chase the emerging grebe, either flying 10–20 yards and diving at it, or racing across the water, paddling furiously with their wings, before dashing alongside the grebe or diving after it if the grebe disappeared again. Sometimes a goosander would fly over, nearly close its wings and dive in head-first, like a gannet in a shallow-angle surface dive. Or it would race across the surface, sinking gradually as it went, leaving a long V-shaped bow wave. Once a female goosander was next to a grebe

with a fish but took no notice: it seemed that the drakes were always the aggressive ones.

Grafham also often scored highly after a gale: perhaps there might be a phalarope or a Leach's petrel. The infamous October 1987 gale deposited rare birds all over the UK, and Grafham received its share. The height of the gale was on a Friday and Grafham had 20 phalaropes, five great skuas, four pomarine skuas and five Sabine's gulls. I couldn't get there until the Saturday, when it was warm, sunny and calm, and I watched a brilliant grey phalarope, a kittiwake, three little gulls, four Sabine's gulls, Sandwich, common, arctic and black terns, and a red-necked grebe. There was a grey phalarope at Paxton, too.

Everything was a bit far out, the Sabine's never coming close, but I spent ages watching them and ended up with pages of notes. The red-necked grebe, however, was certainly close up and exceptional.

But Sunday was different again. Most of the birds were still at Grafham, but now the Sabine's were out over the centre of the lake in a blustery wind, looking really good over rough water. Then I moved around the reservoir to try to get closer and they seemed to have moved: but there they were, on the shore! Three adults and a juvenile. They fed on a strip of sand and even over an adjacent ploughed field and I could watch them at 20 yards for long spells: at times, barely 10. Sometimes, one would come too close for me to focus my binoculars on it. One walked calmly by me, on the beach, within 12 feet. As with subsequent Sabine's, I particularly noticed how beautifully controlled they were in flight, the forked tail twisting sideways and the rather broad-based wings seeming to give perfect poise, delicate and accurate when tilting or dipping to pick up a morsel of food.

Blashford Lakes

Blashford Lakes on the edge of Hampshire is more reminiscent of Little Paxton than, say, Chasewater, being the result of extensive gravel extraction, this time along the Avon valley. The area was an RAF aerodrome in the Second World War, before becoming a motor-racing track for a time, where you might have seen Mike Hawthorn or John Surtees rather than a great crested grebe. It is now largely a popular and well-managed Wildlife Trust nature reserve. It became my local place of choice around 2008.

There are several lakes large and small, with Ibsley Water and Ivy Lake perhaps the principal birdy ones: Ibsley has the main gull roost and that has become a great focus of afternoon activity for much of the year, best watched from hides until chucking-out time. Birdwatching here is, more than at any of my previous 'local patches', more hide-based and it is less easy – impossible mostly – to walk anywhere along the shoreline, although there are plenty of woodland footpaths between the lakes. So it is more a case of sitting comfortably with a telescope and hoping birds come close or settle in view: if not, you can't do much about it. Good for reducing the disturbance, anyway.

There are dense thickets of alder, willow, bramble and so on that look much like Paxton but lack the nightingales. The woods might, though, have a lesser spotted woodpecker – and in summer they are always good for listening to and comparing singing garden warblers and blackcaps. One hide has several well-filled feeders within a few feet, giving quite exceptional views of such bird as siskins, redpolls and great spotted woodpeckers: I have often noted these as 'in-the-hand' style views, as it is possible to examine every detail as if you had caught the birds to ring them.

It is okay for waders (one of my best finds for many years was a long-billed dowitcher on a smaller lake), and good days can bring really very close views

Redpoll and siskin.

indeed of, say, little ringed plovers and snipe in front of a hide. The lakes are good, too, for passing terns, but also for such things as black-necked grebes – which appear quite regularly, especially in spring. A hide looking at a close patch of tall wet vegetation is excellent, giving just fantastic views of bitterns if you try hard enough and often enough. For years there was a regular wintering great white egret, which was ringed and each summer would go home to France; once there were two in the valley.

Wildfowl are always good, with beautiful goldeneyes and goosanders often resting on the shore, showing off their red legs in contrast to the pale yellowish or pink of the drakes' flanks, depending on the lighting conditions. Smews are uncommon but always possible in winter, gadwalls and wigeon often abundant, and there was for a few years a regular ferruginous duck, although it was not always easy to see. Birds of prey of several species are frequent. All in all, plenty to be going on with at this fine reserve, making a decent local patch even though I never quite 'owned it' in the way that I might have in my early days. One spring was especially good for adders, which allowed themselves to be viewed at close range for a few weeks.

In 2012 there were, for some reason not quite explained, exceptional numbers of shovelers: second, I think, only to the Ouse Washes at the time. Numbers at

least equalled those I observed on the Somerset Levels a few years before, which I recorded as the most I had ever encountered.

Sunday 16 December

*Shovelers	635 (apparently a new county record count)
*Pintails	100+
*Wigeon	500–600
Goldeneyes	5–10
Goosanders	c. 5
*Red-crested pochard	1 male (whatever its origin, a superb bird)
*Black-headed gulls	2,000–3,000, some great effects in very strange lights; one with complete summer hood
Yellow-legged gull	1

Great numbers of wildfowl all over the calm lake, apparently after shooting in the valley earlier. Spectacular numbers of shovelers and terrific pintails. (Subsequently found that there were some 25 more shovelers in the 'goosander' bay, too.)

Wildfowl all over the lake keeping absolutely still (just one male wigeon with head tilted back, bill almost vertical) during torrential downpour, the water silvered all over by bouncing drops and hail.

Monday 17 December

**Shovelers	750+ (760)
*Pintails	99+
Goosanders	c. 6
Goldeneyes	5–10
*Black-tailed godwits	500+

After yesterday's phenomenal count, I tried hard to get a really accurate count using my digital counter, but it proved near impossible. Several counts were aborted in the 400s when gulls caused disturbance, but then I got a good settled count of 700 (precisely!) After more disturbance brought many more wigeon and pintails, and some shovelers, out of the hidden 'goosander' bay, I got another good count of 760, but really too many were far away and mixed up, by then, with wigeon etc. – call it 750 anyway. The 'new county record' didn't last long! I thought there were more pintails than I actually managed to count. Nothing really feeding, just drifting, so in bright sunshine, close-up shovelers, pintails, wigeon and teal were just fabulous.

Sunday 27 January 2013

*Shovelers	1,190 or thereabouts, in quite a decent count, well settled although many quite distant – 1,170 plus a few that came up with the pintails
*Pintails	400–500 – not a proper count as most were in a huge flock that flew up from the right hand bay briefly but went back down out of sight – but very obviously far more than I have ever seen here before.

Getting to know the local river

Close to Blashford is a famous fishing river, the beautiful Hampshire Avon. It comes down from the chalk and is often brilliantly clear, but heavy rain will muddy things a lot and, in very wet spells, it will break its banks and spill over onto what used to be regular water meadows. These have scattered low concrete structures, some marked with dates, more than 100 years old, the remains of the sluices and channels that controlled the periodic flow for the benefit of the grazing livestock. The flow of water over the grass can be deep and quick and reveals the clarity far better than the main body of the river. Swans usually feed on the grass but scatter across the floods between clumps of rushes. The fields look good for a corncrake reintroduction, but are usually too quickly mown for silage.

The river is a classic 'Mr Crabtree' river (Bernard Venables's 1949 creation: a great writer and excellent, evocative artist whose mix of essays and cartoon-style strips created a book that sold millions). It is full of deep eddies and dark pools, with overhanging willows and thickets of tall, dense vegetation, shambles of brambles, little patches of mud full of assorted footprints (even some of otters) – a magical one. It could equally be *Wind in the Willows* country. Smaller tributaries are apt to flow in from the chalk and become much bigger in winter (winter-bournes), and are typically stupendously clear.

During my few years in the area from 2007, I witnessed the demise of the local Bewick's swan flock: what had been 100+ each winter in the Avon valley near Ibsley had already dwindled to 20-ish, and by the turn of the 2020s, there were none. The white-fronted goose flock once established there had gone long before. Both are a result of the 'short-stopping' phenomenon, with these wintering wildfowl no longer bothering to come so far west in our milder winters.

Wintering flocks of lapwings seem to have gone the same way, too, although the old water meadows, on the face of it, do not look much changed.

That, I suppose, is one of the reasons for your local patch watching. Apart from putting in the hours to give a better chance of finding something good, you do notice changes over time, even if you cannot always explain them.

Nearby is a little modern-ish sewage farm (not the old-fashioned open type with flat settling beds that used to attract waders). Sewage farms presumably have an abundance of insects and seem to be linked with wintering warblers such as chiffchaffs. In hedgerows alongside this one, I can usually find a wintering chiffchaff or two. One year there were more.

In a nearby field in 2008 were 29 little egrets. (More often I would see up to 10 or so, and maybe a great white egret too. Little egrets have greatly increased UK-wide in recent decades, but here they have gone downhill a lot since I saw these good numbers: even one is now a notable sight along this bit of valley.) There were kingfishers, grey wagtails, 75 or more pied wagtails, redwing flocks in the fields, Cetti's warblers giving their first snatches of song at the end of winter, and long-tailed tits in the hedges with the chiffchaffs. It all seemed to encourage more watching, more searching.

My notes:

> A curious mixture of species. I explored more of the footpath to the north (very wet) – the 29 egrets were excellent, but often very flighty and shy. Although nothing special now in UK terms, this is still the largest number by far that I have seen so far in Britain. All were on the flooded fields by the river in the sewage-works area. I rarely see winter chiffchaffs. These were loosely associated with long-tailed tits, in dense hawthorns and hazel with a lot of ivy. One was classic 'green' with a typical *hweet* call, the others silent and duller, one brown-and-buff but with fine greenish fringes to the flight feathers, another similar but more yellowish visible around the sides of the breast when the bird stretched or the feathers were ruffled by the wind. Often three in view at once, but I think there were four.
>
> Only one spray, but already bright green hawthorn leaves.

I began to think that one of the chiffchaffs was a Siberian chiffchaff (variably considered a subspecies of chiffchaff or a species in its own right – you will run into this problem a lot). Once I decided to investigate properly, it had gone (you will run into that a lot, too).

Nearby in the grassy meadows there were some Bewick's swans:

> Bewick's swans are really beautiful things – excellent views of them.
>
> Floodwater has receded a little, but still very extensive: water on the fields flowing quite quickly westwards across the valley to the smaller river the other side – to the south, still impossible to tell where the river channels are.
>
> The rivers are much clearer than before this winter, but I arrived too late in the afternoon to see any fish really. A superb sunset, though.

Bewick's swans are indeed beautiful birds. Numbers here declined year on year but I still usually managed a few good, close views. Once, one caught my eye right away and proved to be a whooper swan, much rarer here. As well as these and the usual mute swans (sometimes hundreds on the fields) there would (and still will) sometimes be a black swan. Black swan is a very different species, from Australasia, here just 'escapes' of course, but they might eventually become established: a few pairs are breeding in England now. One odd thing I saw was a black swan together with a pair of mute swans and their cygnets, so close as to be 'rubbing shoulders' with the parent pair. Had it been another mute swan it would assuredly have been given short shrift and driven away. Black swans have

Mute swan.

become fairly frequent, but there may just be a handful of long-lived individuals that exaggerate the apparent numbers.

Fish give a break from birds, with just as much entertainment, often in great surroundings. They are beautiful things to watch, but sadly neglected by naturalists. I've deleted practically everything about butterflies, snakes and wild flowers from this book (they feature strongly in my notebooks), but I must keep just a little bit about fish:

> In the little side stream by the road near the mill, with much clearer water, 3–4 *dace, 3–4 small chub, 3–4 perch, scores of minnows, a gudgeon, a fair-sized pike and several *brown trout, one quite big, all beautiful, coppery-brown above, with big, buff-ringed black spots and bold black spots on the dorsal fin. Whether the water or the light, or both, something imparted a cerulean blue look to the flanks (from above, less so from the side), the base of the fins and, brighter, around the mouth – even the pike seemed tinged blue.
>
> *Chub: 10–15 'big' ones of maybe 5–6 lbs plus, and many smaller fish; scores of *dace; a few bleak?; one perch; one brown trout plus one trout/salmon parr; ruffe?; 100+ *gudgeon, maybe many more. Lovely views of dace – often dace and small chub side by side. Middle-sized chub the best, looking bronzy, with red lower fins; the big ones were drab grey by comparison, with grey-green fins, often without any hint of red – but I could see inside the gills to patches of bright red-pink.
>
> Masses of gudgeon almost pressed to the bottom, some swimming more freely – I could often see the barbules. From above they seem to have a pale spinal line broken by dark bars, and the sides are almost striped on a broad body, with bumps around the eyes on a long head – odd-looking things. Some more evenly patterned ones may have been ruffes, swimming more freely above the bottom.
>
> *Grayling c. 10–12, seen remarkably well in very clear water. Best views of grayling ever, I should think, at least to be able to appreciate all the colours and patterns – some beautifully marked fish, some much more extensively black-spotted than others. Most had whitish mouths ahead of a band of lilac-blue, sheen of lilac and violet, and pelvic fins marked with bands of pale turquoise and chocolate brown.
>
> At about 11 a.m., in hazy sunshine, a small group of black-headed gulls stood in a riverside field (east of the river, north of the road) and suddenly

began to call and flew up over the field, out of sight behind a tree. As the calls continued, they drifted back but clearly over something in the field – amazingly, a superb OTTER came bounding across the field, long tail trailing, and belly-flopped into the riverside vegetation. The gulls continued to follow and a line of bubbles came my way – then the otter appeared on the surface, and swam downstream, towards me, under the other end of the bridge on which I stood, and then on as far as I could see to the next river bend. The gulls continued to follow, calling, and the mute swans, moorhens and others took evasive action!

The otter looked a big and formidable animal. When it first resurfaced it showed full length for a while, then began a regular sequence for the rest of the time I watched it, 'porpoising', but almost more like a dolphin – or a whale! – even with something of a human look about it. It created a bow wave on the surface each time it appeared, just bubbles while it was submerged. The head appeared for a few seconds, then lunged upwards as if taking a great gulp of air, rolled forwards and disappeared as the bulging back, then tapered tail, followed. It was going downstream and seeming to be in such a powerful hurry, but maybe made less speed than I imagined it might. Brilliant.

Looking at redpolls

The bird feeders very close to one of the hides at Blashford Lakes give exceedingly good views of many birds; the sudden flurried arrival of a great spotted woodpecker is astonishing! Mealy redpolls (or common redpolls as they are now usually called: those from the continent rather than the 'lesser redpolls' breeding in the UK) are rare there, but one year we saw a few. Lessers are the small dark, buffy ones, mealies the larger, paler ones with whiter wing-bars and rumps (but the deeper buff of a lesser redpoll's wing-bars fades whiter as the feathers age). Some extracts from my notes:

Redpolls extremely variable. One or two small, dark, very deep, rich, bright tan/tawny on all 'buff' parts, with forehead/lores/bib 'black' all suffused with tawny, wing-bars deep tawny, red crown small; others paler, buffer, whiter below, with all 'buff' parts including wing-bar and supercilium paler clear buff, lores and bib blacker; others a bit paler/more contrasty again, one or two with long whitish stripe on back, lores/mask/bib particularly large and black. One male quite nicely pink with pink rump, another

> extensively red-pink on chest, pink on rump. One at least looked much like the slightly larger mealy, but less grey, with whitish wing-bars and whitish edges on back feathers (but no obvious long line), rump not seen; two long black flank stripes. One **mealy** large, long-billed (relatively), pale areas of superciliary etc. pale greyish or very cold whitish-buff; smallish red cap; dusky grey upper and rear edge to pale cheek with flecks of pale pink; back feathers neatly and finely edged whitish but no long lines; extensive black mask/lores/bib; some very pale pink on upper chest; long, scalloped median-covert bar white (usually more hidden and buff/tawny on lessers); greater-covert bar broad and white. Rump nearly always hidden except for a small triangle of white between wingtips, but briefly showed much broader/deeper white area with mottles at least on upper edge.

In other words, from small and dark and toffee/tan coloured to big and pale with white rump, almost a complete gradation. Very tricky. Was the 'nearly mealy' a mealy, too?

> Extremely good views of the redpolls, although more restricted to the feeders than yesterday, which means they tend to be half-hidden, or turned the wrong way, and hardly move, so you can't see the rump, or back, or whatever you particularly want to check. A 'male' mealy redpoll was reported earlier. I saw this one (which I had not seen before), then the same bird as yesterday and a bird that was probably a third mealy (which I think I also saw yesterday).
>
> So there was a 'good' mealy with whitish fringes on the back, and no pink on the breast (seen quite briefly); a 'good' mealy with extensive pink on the chest as well as a few flecks on the cheek, with a very good head/nape, but a more buffy back and a little buff on the outermost part of the greater-covert bar; rump seen less well but seemingly similar with extensive white (this one seen much longer and more often). Both have surprisingly long dark flank stripes. Then there is a third bird (at least), also looking biggish, quite large-headed, with a good head and nape as on the others, but a generally buffer appearance overall above; no pink except minute specks under the eye; flank stripes more broken, shorter and less black.
>
> Lessers very variable – some stunning males very extensively red and pink; one 'toffee brown' bird (like juvenile?) with rich tawny wing-bars also extensively red on the breast. There seems to be little clear definition of age/sex and often there are two birds, alike in pattern, one very pale and buff, the other dark and rich tawny, side by side. Hard to work them out.

> Superb views of the redpolls: really great, prolonged views of mealies. The pink-breasted male preened and showed its rump extremely well, fluffed out, white with pale grey triangular streaks on the upper edge, slightly darker and browner triangular streaks on the lower edge, and a faint pink wash across the central white area.

Making notes was always enjoyable, sometimes a bit of an effort when it came to putting them into a fair copy. There was a fine day, briefly alluded to before, when I saw more good redpolls. It was 11 February 1996: Norfolk. I filled pages and pages with notes on waxwings (well, you need to get them right), a serin, flocks of snow buntings, a little auk, even 8,000 pink-footed geese and a pale-bellied brent. Near Cromer there were just two or three lesser redpolls, about five common or mealy redpolls and one or two arctic redpolls.

> I had brief fairly distant views, then a closer view of an arctic redpoll, including good views of the completely clear dull white rump and completely unmarked white undertail coverts, seen well from below.

Then at Langham, near Blakeney, there were 40 redpolls that mostly seemed to be excellent mealies (something I was not very familiar with), plus at least four male arctic redpolls and perhaps some females/juveniles. There was evidently some difficulty with preconceived ideas gained from identification books or papers:

> Brilliant – amazing views of redpolls at the closest limit of both binocular and telescope focus – barbule-by-barbule detail! The closest views included arctics. I had not really grasped the full range of variation in lessers (few) or mealies (one really vivid male) so I was not really making progress in determining real or consistent differences from arctic. For example, I thought the head shape of arctic was distinctive [*I assume here that I meant I had assumed that in advance*] but it is too easy to believe it to be so – some mealies looked very similar. Or were they female/immature arctics? Some mealies showed pale rumps, but the arctics showed big areas of clear white (but a 'dull' white, not a vivid bullfinch white). Mealies also had rough feathery 'shorts' on the thighs [*another feature sometimes proposed for arctic*], although arctics on the ground showed this feature particularly well.
>
> Arctics looked broad, long, rather bulky and upright when perched with a 'pushed in' face but a rather steep forehead up to a peak behind the eye on a rather sharply domed crown. Their bills looked tiny, straight-edged, fine and very short; their eyes really small and close-set. On the ground the head

> looked flatter, with a smooth curve from the bill over the head and nape to the back, so the front of the back made a smooth whale-back shape. One at least had a very obviously deeply forked tail with very sharp corners.
>
> They could look quite grey, but not always obviously so, nor necessarily whitish on the back, but in general they looked pale, colder, greyer than a mealy, greyer on the nape and crown area with a small vivid scarlet cap not reaching the bill (mixed with blackish when wet); white superciliary and line in front of red cap. Black around bill, eye and on small bib. Bill yellow with blackish ridge and tiny black-brown tip. Greater coverts had broad white tips extending in a triangle along the edge of the outer web, giving a 'sawtooth' upper edge. Broad white tertial edges...

Well, having reached the tertial edges, perhaps it is best to stop: the notes went on for many more lines, detailing every feather tract. The value of such things is not entirely obvious, but I am glad I always looked so closely, so hard, so determined to get things right, rather than just superficially clock the white rump and have done with it.

Notes such as these were not at all obsessive: they were made if the subject required, or demanded, them. The vast majority of my days out had records but no further descriptive notes; maybe a line about the day, the weather or whatever. Of course, detailed descriptions may not be repeated once a species becomes more familiar, and no-one suggests writing copious notes on commoner birds – but still, making detailed descriptions of new or difficult birds is a good habit to get into.

WONDERFUL WALES – CROESO I GYMRU

While I can add nothing that has not been said better by others to reinforce the attractions of Wales, the country has given a great deal to me and has undoubtedly enriched my life. Even though it is a long while since I lived there, my affection and affinity for things Welsh remains. Perhaps my Welsh student friends in the rugby season excepted.

Roger Lovegrove and Graham Williams in the RSPB Wales HQ, in Newtown, employed a number of people on short contracts and I was enormously grateful to be numbered amongst them: people like Tim Inskipp, Tim Cleeves, Martin Davies (of Birdfair fame) and Keith Vinicombe. Where would I be now... Roger gave me such a big break and later pointed me towards a permanent job at the RSPB.

My first job, in 1976, was to survey the birds of some of the best of the mid/north Wales hills and moors, so off I went to Llangynog with my little tent (I still remember the smell of the blue groundsheet, punctured by a spike of dead bracken on my first night). Later, nights were spent in assorted caravans, even an empty school where I had to use the washrooms and toilets made for tiny infants. The local village had a café where I managed to extract unprecedented numbers of cups of tea from one teabag and, memorably, one of the villagers said to me 'A beautiful day!' in the pouring rain, but was quite right. It was. I was employed on other things for the winter, then had another season outdoors the following summer, on hills farther south in Radnor, based near Penybont.

Naturally my diaries mushroomed: day after day, birds, butterflies and plants counted and recorded, mapped and written up in duplicate or triplicate. I recorded butterflies on special little cards, mapped the best of the remaining heather moors, estimated the proportions of constituent tree species in local woods – far, far too much to do more than sample very briefly here. But a flavour, maybe, will be of interest. What would these hills have to offer now, I wonder? I can't help but think they would be much diminished. I learned recently, for instance, that curlews have almost gone from the Berwyns: just one pair bred at the RSPB's huge Lake Vyrnwy reserve by 2018. How can such an inescapable and indispensable part of the Welsh moors, the voice of the hills, have gone, just like that?

There were many special places with interesting and important birds of prey. At the time, most were rare: peregrines had barely begun to recover their numbers, hen harriers were secrets of remote moors, if anyone told you where red kites nested he'd have to kill you... I well remember the eminent Welsh writer and broadcaster Wynford Vaughan-Thomas saying on the radio (or television? – memory not so good after all) that he knew a place where he could see red kites, but wild horses wouldn't drag it from him. Merlins were probably at a higher level than they have been since, though, to the delight of RSPB assistant regional officer, Graham Williams.

There are several things to say before getting into details of particular days and moments. First, this gives an idea of the results of fairly intensive survey work, or at least a lot of hours put in on the hills. If you are a beginner, you may want to learn about a broader range of birds before you concentrate on a particular group or the birds of a particular area or habitat; that can come later. But then again, if your local area is like this, why not now? Purposeful birdwatching. Second, you may not, anyway, be able to go to an area of upland moors like this; if you live in parts of Scotland, Wales or Ireland, you might, but mostly not. And third, importantly, I was armed with a 'Schedule 1' licence, meaning I was able to look closely at some rare species: not that the birds would know, so I was still extremely careful about what I was doing, and did my best to create the least possible disturbance. I was also aware of drawing attention to rare birds but, to be fair, there were almost no other people out on the hills back then to be interested in what I was looking at. I was issued with a compass and a whistle, but really was out on my own far from any roads, very often, with no such luxuries as a mobile phone. Once I did fall down a deep, heather-covered hole in an eroded peat bog, but I was lucky enough to be able to crawl out.

One small valley had a cliff on one side and an area of bracken with scattered trees and scrubby larch on the other. I sat on a bank below the cliffs sometimes, with peregrines occasionally calling above me, redstarts nearby, grey wagtails and common sandpipers on the river below and black grouse displaying over the far side. I could hear the cooing and 'sneezing' of the grouse more than half a mile away and just make out their dark shapes. When I walked over that slope I saw a couple of black grouse well and then heard a falcon calling: the first of two merlins. They were later flying together over a ridge, with steep, tussocky heather slopes and scattered small crags. Then a male hen harrier rose above the ridge and turned away to disappear beyond it. I clambered up, looked towards where he had been heading and saw instead a female. I didn't think they would be nesting nearby, but it did look a good spot for the merlins. The variety of birds matched the rich textures, shapes and colours of the landscape. Almost 50 years on, I look back at all that and wish I could still see even half of it.

Cuckoos were calling all the time in some places but it was not until late April that I actually saw one. Dippers fed their young in a nest beneath a road bridge, on a stream where there were also common sandpipers. I particularly noted that willow warblers were everywhere and singing brilliantly (would that they were still as common) and chaffinches were equally frequent. Not much beats the fresh spring song of a newly arrived willow warbler in a bright green, fragrant birch tree.

On the crags next day there was a lot more peregrine activity, and birds of prey, together with ravens, gave some wonderful moments in combination. Four kestrels, a sparrowhawk, three buzzards, two of the splendid peregrines, a couple of ravens. These were flying around the cliff, momentarily settling before sweeping up again and circling at a distance (I was watching from quite a long way downhill). Both were out of sight at the moment a carrion crow appeared, and then the female reappeared and dived at the surprised crow's head. Peregrine calls are quite varied: often described, as most falcons' calls are, as a *kee-kee-kee* but I noted down a gull-like, quivering or wavering *keee-ee-ee*, thin, nasal *eep eep* notes, a throaty, lesser black-backed gull-like *kyoww* and, from the male, a sharper *kyow-yo-yo*. Long, harsh, grating calls and chesty, hissing *hair-hairr-hairr* notes are also typical of peregrines.

Next up came a valley with a good number of wheatears and, especially good, because they were (and are) scarce and not something I saw every day, a pair of ring ouzels, as well as tree pipit, curlew, buzzard… nice stuff, always worth a long look.

The next day, continuing coverage on an adjacent patch, my notes include five or six singing tree pipits, a redstart, yellow wagtail, several wheatears, two pairs of ring ouzels, ravens, several red grouse, stock doves (they like uplands, with quarries where they nest), curlews (still good numbers), three kestrels, two peregrines, two merlins, two hen harriers, dipper, grey wagtail and twite. The twite was noted as 'heard', which was a bit confident perhaps for a bird so very scarce in Wales, but I was fully tuned in to twites then, with plenty of opportunities to see my 'local' flock in Staffordshire each winter.

It was a simply marvellous place: all these things singing away while birds of prey performed overhead, and a landscape to dream about. A brilliant male hen harrier looked ghostly pale against dark heather, from which it flushed (and dived at) a merlin, but the little male merlin shot away, turned and came back to dive violently at the harrier. Both settled on small poles, before the merlin had another go. As I approached, the male merlin was again on a post but a female rose up and dived at him, too. The male merlin seemed a bit upset by all of this, shot out after a meadow pipit, missed it and, seemingly in frustration, caught a moth instead. There followed more interaction between hen harrier and merlin, and other times when neither reacted to the presence of the other.

Now a peregrine appeared and seemed to want to show the others 'how to do it'. It just circled, gaining height for several minutes without so much as a flicker of its wings, until it was several hundred feet above a broad cliff. As I looked with binoculars I saw a speck even far above that; I checked, and it was invisible to the naked eye. It was a surprise: the peregrine came down and the mystery bird from the heavens above dived at it, none other than that same aggressive male hen harrier. These two had a real battle, or pretend battle anyway, circling and diving, sometimes almost confusing as to which was which as it was not always obvious that the harrier was the larger bird in the rough and tumble. A second peregrine appeared then, clearly a big female, and it closed the wingtips back to

Sketches of merlin, hen harrier and peregrine.

the tail, taking on a broad, heavy 'teardrop shape' before plunging towards the cliff, checking a couple of times before diving in to a hidden ledge.

On other days, after I had wandered far and wide and needed a rest, I was in the same general area and enjoyed the 'ordinary' birds of mid-Wales such as redstarts, wood warblers and pied flycatchers, the great trio of the western oakwoods (although wood warblers would also sing from little copses of larch and redstarts also spilled out into the bushy fringes blending across rocky slopes). Wood warblers and redstarts also nest in other places across midland and eastern Britain, scattered spots in the New Forest or the Chilterns for instance, but pied flycatchers hardly do so. All three are really most closely associated with the west, for reasons that are quite hard to define.

Wood warblers have a sharp, ticking note that accelerates into the classic silvery trill song, but these were varying it quite a bit: sometimes all ticks, sometimes no ticks and all trill; some trills high, some low, sometimes a low trill that suddenly changed midstream to a higher one. As ever, such variations are far too much to cover in an identification guide, but it becomes more important in a European guide with Bonelli's warblers included. They sound rather similar but with a different quality of trill, without the ticks. If you have no room to mention the fact that wood warblers can also drop the ticks, sometimes, you don't properly highlight the potential for confusion and the book is less comprehensive than it should be.

There were remarkable numbers of whinchats on the fringes of the hills, mostly in bracken, and a few pairs of stonechats – which were less expected. They tended to be more coastal, and inland pairs came and went according to the severity of the winter weather. Some very high, boggy pools had little colonies of black-headed gulls, just on eggs, the nests each tucked into a little tussock of rush. Redshanks, snipe, curlews and lapwings all bred on the lower commons, lapwings also in nearby fields, and I was finding nests with eggs and later watching the development of the chicks from balls of down to fledglings. It was a remarkably rich area. One day I had counted 26 singing whinchats on one hill – amazing – virtually all in bracken and none in heather, bilberry or rushes. There were ravens about most of the time. These were almost honorary birds of prey, often soaring, and with their big nests on ledges or in tall trees.

> Two were fighting crows and creating a chorus of popping, croaking, rattling and machine-gun calls and flew around for long periods 'duetting', flying almost together, turning, soaring, diving and rolling in unison, like a bird and its reflection, frequently with head and neck feathers raised to create a hugely top-heavy impression. Then they became extremely agitated, flying close

In mid-Wales I found very many whinchats on slopes of bracken, heather and scattered hawthorns. They disappeared once such places were changed to short grass for sheep.

> overhead, calling constantly; one chased a swallow several times. I noticed the nest, with young just visible, in a slightly odd site – 40 or 50 feet up in the central fork of a large sycamore, an isolated tree in a long, low hedgerow. It was a large, very deep nest. Later, ravens were noted all over the hills in ones and twos, then the same noisy ones again. I scanned the trees for the nest but found a buzzard brooding, instead, then one or two more ravens flew up from the valley bottom. Young ones perhaps? As I watched a great flock of birds issued from the trees, and incredibly all were ravens. They gathered into a very noisy flock and all soared, dived and rolled around in a wheeling mass: at least 55 together. The flock eventually split up, some going off one way, others in a different direction, while about 20 settled on a heathery slope.

This was early in May. On 19 May I saw another congregation of 65, as well as another nest. In Scotland I have seen up to 110 ravens cavorting together, and even recently in Hampshire a similar number, something entirely unheard of and unexpected only a couple of decades ago.

Peregrines were always exciting. There were more than had been expected. One day there were peregrines on a cliff, two or three merlins, a kestrel, two buzzards, two or three hen harriers (let alone the dippers, ring ouzels, tree pipits, pied flycatchers, whinchats...). The peregrines had a downy chick on a ledge, in an old raven's nest, a huge stick pile. The female gave magnificent views while the male flew around screaming abuse, turning, diving, stalling and generally showing his disapproval of my presence in the area. Later, the view of the splendid male, his colours and pattern that little bit extra-sharp, his feet brilliant yellow, standing on a raven's nest on a huge crag, cuddling up to his downy chick, was a memorable one. I had found a merlin's nest the previous year but it was not occupied, although there were various plucking posts around, mostly on shooting butts. I was happy not to be around come August. They were mostly decorated with meadow pipit feathers and the wings of emperor moths. Meanwhile, a male hen harrier brought food and the female appeared for the 'food pass', swiftly turning over to snatch the prey, all over too quickly to appreciate.

A few days later it was an exceptional day all round. A little owl was a surprise start. I found two dead ravens by a footpath: suspected to be poisoned. A local landowner had told me that it was good practice to put out poisoned eggs, as doing this, and killing a few harriers, would remove the 'strain' of egg-eating harriers and everyone would be happy. He wondered why black grouse had declined, when not so long ago the estate could account for a couple of dozen cocks shot each year. But he did offer me a dilapidated and disused hut on the moors in which I could set up camp for a few nights. Golden plovers sang and displayed; curlews called everywhere. Curlews, which now worry everyone involved in bird conservation: what is it about curlews, with various species worldwide in such a parlous state? Some already gone. Cuckoos created a wonderfully bouncy, soft chorus. A tiny pool had at least 19 black-headed gull nests with eggs, mostly typically mottled olive and matt black, some basically pale blue.

A female hen harrier chattered above a man with a gun – I very rarely saw people up here, and the gun suggested a gamekeeper. Maybe he was the one who had been convicted recently of shooting over the heads of a Sunday school outing walking up the hill. The male harrier appeared nearby and both birds remained in the area, calling occasionally. I walked towards a rock on which I had earlier found a peregrine nest with eggs, which should have hatched long ago by now. It was seemingly vacant. I climbed to the top: no sound, no sign of anything. Worried, I peered over the top and, with a great shriek, the female flew off six feet from my face. Still three eggs... they never did hatch.

The peregrine flew around, and so did a hen harrier. I had a screaming peregrine and a chattering hen harrier, one each side, as I walked away. A male harrier appeared and dived at the peregrine: these two, plus the original female harrier, continued sparring for several minutes. The female harrier came within 30–50 yards but the male was the bolder of the two (unusually), sailing right overhead, several times pulling his wingtips back into a 'teardrop' shape and diving to within 15 feet. He was, against the blue sky in bright sun, a fantastic sight, his yellow eyes blazing in the sun. Sometimes he just hung motionless above me at about 30 or 40 feet. The trio reconvened and more sparring ensued.

At a different hen harrier nest, it was the female that was, apparently typically, the more aggressive, the male more timid. That pair had small chicks, still showing a little white egg-tooth on the tip of the beak, while another nearby had six bright blue-white eggs. The female, doing the incubating, was quite heavily in moult, the male still neat and immaculate.

By early July the female peregrine was still incubating. She was less vociferous; the eggs looked a little sad. Harriers dived even closer, 10 feet from my head. And the day was topped off by wonderful views of a mole. It crossed the road in front of me, and let me watch it for a few minutes.

A few days later I was at a merlin nest, with apparently two adult males and four juveniles in attendance... There were various decapitated birds in the empty nest: meadow pipit, skylark, great tit. And then there was a hen harrier nest. Wow! A small platform of heather twigs with a pad of heather, rush and grass, about 18 inches above the ground in three-foot-deep heather. Five marvellous young sat back on their tails and glared at me. I quickly looked and then took off, leaving them to it.

The next day, another cliff, two peregrines with two large young: this was how it *should* be.

By 13 July, still three hen harriers near the nest but two found dead nearby. One juvenile drew blood from my finger until I put my boot in the way and it happily clawed at that. All three had dark brown eyes, so presumably female: I noted that their tongues were bright pink, with a green tip, just like merlin chicks.

In a different area, I scrambled up a heathery gully and saw a fine cuckoo and a pair of ring ouzels. In a patch of old larches near the top there were droppings and pellets indicating the presence of owls, presumed long-eared owls, but I couldn't see them. Suddenly, there was a male merlin chasing a crow. Later a female flew out from the same place, brilliant in bright sun. There was an old crow's nest in the larch trees, so I walked back 150 yards and lay down in the heather. I couldn't quite see into the nest but the male returned, calling loudly: an absolute cracker, bright in the sunshine, blue-grey and rusty-orange with vivid yellow legs. Previously I had heard what I assumed were 'display' notes, *kee kee kee*; now the male gave sharper whistles, *wit-wit-it-it-it*. He moved to the nest and settled down as if incubating. The female appeared, with longer call-notes – a whining *wheeeet-weee weee weee wee wee*, distracting me from the male, which left without my noticing. Both, in fact, disappeared, so I moved off too.

It was mid-May before I returned, and immediately both merlins flew up on my approach, but it was a pair of crows they were after, not me. The male returned with loud, fast chittering notes, the female was silent. He called the same later while chasing some ravens: male merlins are nothing if not brave (I have watched one chase a golden eagle). As he circled he looked back at me, head raised over his shoulder, wings flicked downwards in a cuckoo-like action. In his aggressive chases, he was quick, agile, sometimes fluttering but mostly fast and dashing, as merlins usually are. The female was later on the nest, the male on a nearby post.

In the middle of June, I returned: the female was still on the nest. This was now more than seven weeks since my first visit. Was she brooding chicks? It was cold, windy, so seemed too risky to disturb the bird and I left. But the male didn't leave me alone: he was a spirited creature. He flew around in low, wide circles, calling all the time, on each circuit making a wild, diving swerve over the nest tree. In between times, he perched on posts up to 75 yards away, or as close as 35. In the telescope, he completely filled the field of view. Again, he sometimes looked cuckoo-like, or flew with the normal dashing effect. Also, he used a flight action typical of a merlin homing in on prey, with the wingtips bent back and flicked in and out, like a thrush. Hobbies do something similar.

By my next visit in July it was clear that things had progressed. Piles of droppings and pellets around the nest seemed to indicate a successful outcome. Was this the female? No, the male... no... a juvenile! Now another. Probably three youngsters, in fact, all flying well close to the nest. One hovered above me, about 15 feet over my head. I didn't see the adult female, but the male had a different call, looked more rakish and narrow-winged, smaller than the juveniles, although it seemed odd that all three might be large females. They called more nasal, drawn out, lower notes, variable *eee-eee-eee* or *kair kair kair* or *ee eeya eeeya eeya eeeeeeya*. The plumages were confusing as they looked rather grey, greyer than the female, yet seemed to be big and broad-winged. Many years later when writing the WildGuides identification books, I found female/juvenile merlins particularly difficult, with many debated photographs. Now, rather than being nervous about being too close and backing off, I was able to sit and enjoy the family at close range, for long spells, taking loads of descriptive notes.

Many birds gave me hours of delight and fascination in my RSPB survey days. Anyone living in the uplands will know more about them than I do, but as someone from a lowland area, I found some species particularly appealing.

> Ring ouzels: a side gully with steep banks, one with tussocky heather, the other dense dead bracken. A pair were on the heather, calling loudly. The female carried food, so I sat well back and watched from 100 yards or so. Blackbirds present in the same gully, a male often only three or four yards from a ring ouzel. Female ring ouzel twice bustled the blackbird away. Female very dingy with a poorly marked gorget, male on the other hand as immaculate as any I've seen, with a huge, white gorget and very black head and underparts. Bill bright lemon-yellow in contrast with the blackbird's deeper orange-yellow. Eventually the female flew to the bracken and stopped calling, reappeared, still carrying food, then returned to the heather. Quickly went back to the bracken and out of sight. I walked over and the bird came up from 2–3 yards... A large, grassy nest with fine grass lining, very well hidden in a deep canopy of dead bracken near the top of a 10-foot bank. One egg and three small, newly hatched but lively chicks. I left as quickly as possible.

Ring ouzels in fresh plumage have more or less obvious pale fringes to the feathers, even on the head and underparts, but by this date, mid-May, these would have worn off to create this especially black and smooth plumage on the male. The wing feathers, however, are more broadly edged pale and remain as a characteristic paler panel, unlike any blackbird (which merely has a paler impression towards the wingtip, against the light).

Grasshopper warblers would have been just as easy to find back home, but they are never 'easy', so any encounter was a notable one:

> Grasshopper warblers in an unusual and attractive spot. Male sang first from a very small clearing on the edge of a deciduous wood, with short bracken, bluebells, red campion and broom. Then crossed an adjacent road to sing openly from the top of a short, layered hazel hedge. On this side of the road was a six-foot-wide sloping bank, with the hedge below, then a meadow and stream. The bank was grassy with scattered umbellifers, meadowsweet and a fine show of red campion and bluebells in full flower. Sharp, metallic calls drew attention to a second bird in the hedge, then both were watched at close range running about along the layered hazel poles.

Grasshopper warblers might be expected in wider grassy spaces, with grass growing up through brambles and hawthorns, in little dense tangles from which the male might sing; they also occupy the edges of reeds and similar tall waterside vegetation, or clumps of rushes on wet heaths and low moors (such as Dartmoor). Sometimes you can watch a singing bird closely, if you are very quiet and cautious. Nowadays, however, as with so many others, these birds are scarce and difficult to find in many areas where once they were more frequent. The 'sharp, metallic calls' do have a particular quality about them, but you would be pushed to remember them the next time you hear one a year or two later. It is not something that most of us hear very often. Some birds are around all the time and call as often as you like, while others will be heard once in a while, or very occasionally, and hardly give us a chance to get really familiar with them.

Learning more about peregrines

Seeing a lot of peregrines in my days in mid-Wales taught me their shapes and behaviours: here are some 1970s ideas of flight silhouettes, which aren't too bad. Females are bigger, bulkier, broader- and blunter-winged than males, but it is not always as obvious as might be thought. They sometimes recalled fulmars, sometimes ravens. Females often have a little 'step' on the trailing edge close to the wingtip, presumably a temporary effect of primary moult.

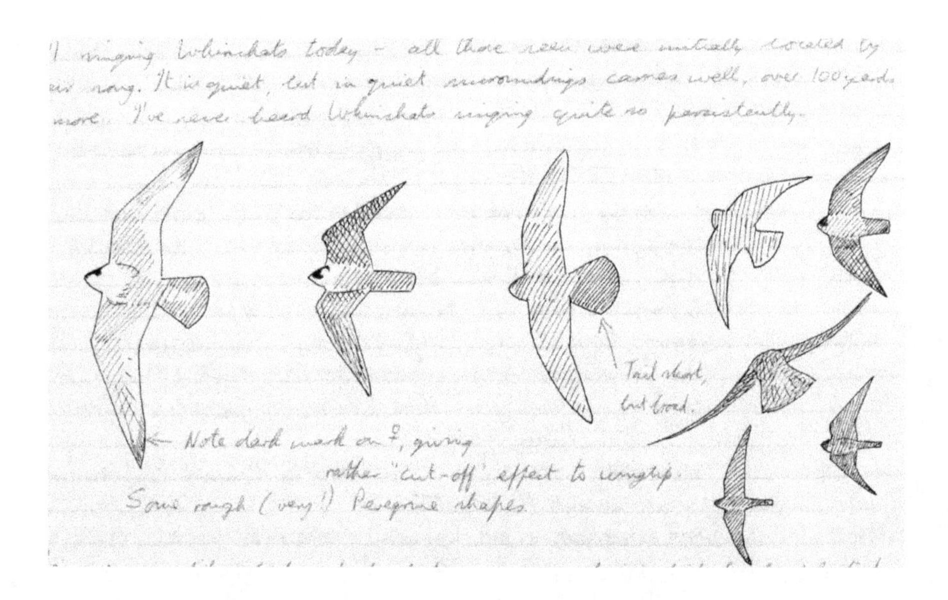

Now and then, in the 2020s, I see a peregrine over my house:

> Mid-morning on an amazingly hot day, I saw what looked like a 'big' peregrine circle up from just to the northwest of the house, quite close. It circled up and up and drifted in various directions, then moved off fast to the south: but it dived and harried another bird that turned out to be a markedly bigger peregrine still, clearly a female, carrying prey. I watched

> the female for some time as it circled quite unconcerned, then switched back to the male – the female drifted off southwest. Briefly the male flew close to it, with a series of odd, deep, quite quick sweeping beats, very like a common tern action. Then the male, with no wingbeats at all, rose up and circled, drifting west, reaching a very great height; it came back lower and closer, then rose again and drifted around for another 10 minutes – beautiful views. Then it took a direct line westwards, with a couple of little tremors of the wings to make minor adjustments, before diving down and closing its wings gradually, entering a fantastic closed-winged, very fast stoop out of sight.

On a visit to South Stack on Anglesey I had excellent views of peregrines:

> I stood by a small stone hut on top of the cliffs looking towards North Stack when a peregrine appeared, low down, coming towards me: it dived at a herring gull for the fun of it, twisted up and came zipping past me at very close range, perhaps as close as I've ever seen a peregrine. It was a big, heavy, muscular adult female. As it reached the top of the cliff steps area, it simply turned into the wind, with wings fully spread, and hung there, just gradually gaining height but almost fixed over a point on the ground – hardly moving a muscle except to turn its head sideways. It was followed by another, clearly the male, which went a little beyond it and then did the same thing. The two of them just gradually gained height without a wingbeat and hung there for many minutes as I lay on my back and watched – brilliant. Eventually the female swung away and made off, and a few minutes later the male did the same. Fantastic views.

It is unusual to see a peregrine in full-stoop at breakneck speed, and uncommon to see a kill. I have on a few occasions heard the remarkable rush of a stooping peregrine, once in the streets of a small hilltop Spanish village, and another time in very different surroundings on the cliffs of Portland Bill:

> Excellent views of a large female peregrine. Later as I was on the clifftop I heard a sudden terrific loud rush and saw two birds, at least, zigzagging down in a fantastic fast rush out of sight under the cliffs beside me – and a peregrine sweeping up and turning, shaking small feathers loose from its feet. It disappeared behind the cliff, too, and then a few white feathers drifted up.

After my summer season on the Berwyns I had an even shorter-term contract based near Mold in Clwyd. It was the late summer of *Dancing Queen*...

background music to my memories. On odd days off I could visit the coast, especially Point of Ayr on the end of the Dee estuary. Here's Friday 13th, not so unlucky, in August 1976:

> Waders roosting at long range and not properly watched: I concentrated on closer gulls and terns.
>
> I don't remember ever seeing so many common (2,000–2,500) or little (100–200) terns. As the tide rose, groups of terns collected on the north-eastern sand spit and muddy shores south towards Talacre mines. In addition, a few hundred fished off the point, well offshore, and common, little and Sandwich terns were passing over the beach to and from this feeding area all the time. I counted up to 50 Sandwich and at one stage saw 78 little terns together. The total could easily have been more than 200. Of 60+ watched carefully, just 20 were juveniles. This far exceeds the local breeding population. Frequently hundreds of larger terns would fly in to settle by a stream or much closer on the beach, and while I could easily identify the common terns, I failed to see a single arctic. However, I picked up a pale-winged tern just as it settled amongst 600–800 commons. On the ground it looked very pale and the telescope quickly revealed a beautiful adult roseate tern. As I watched it, I became aware that next to it there was also a juvenile. As I watched that, the adult became 'lost', presumably hidden behind another bird in the crush. As I watched and made notes, a low-flying aircraft disturbed the flock and I failed to see the roseates in flight. Half an hour later I relocated them: the juvenile walked up to the adult and begged for food.

I took a couple of pages of notes and made sketches, eager to learn this rare and elegant tern, of which I had little direct experience.

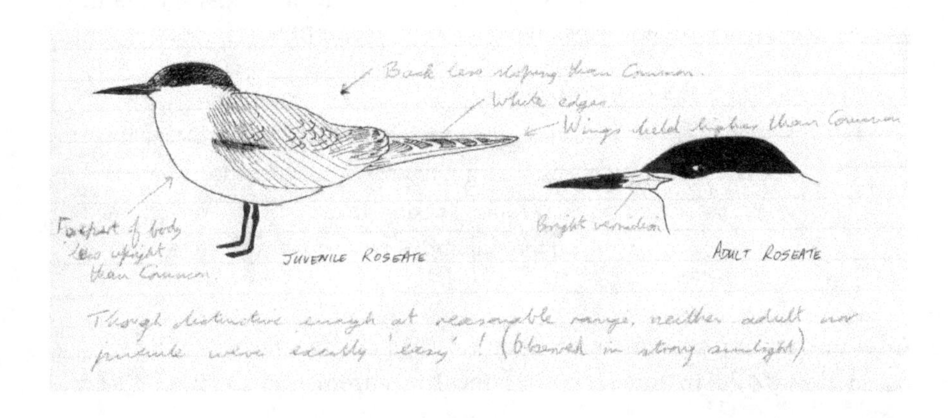

SUPERLATIVE SCOTLAND

Only once or twice have I travelled to Scotland on a 'birdwatching trip' as such: winter visits to see a few rarities (king eider, surf scoter) and some of Scotland's 'specials' (capercaillie, black grouse, ptarmigan, golden eagle). On one of those, my long-unread diary reminds me, we stopped for the night near a railway station at Blair Atholl and I spent an uncomfortable night in the car. Eric Phillips decided he would sleep in the station waiting room. The next morning when he chatted to the station master, he was told that, if he had enjoyed a good night, he was a lucky man, as that room was known to be occupied by a none-too-friendly ghost.

For several years in succession, though, in my school/student years in the 1960s and 70s, we went on family holidays and, courtesy of kind uncles and aunts, I was often lucky enough to get four weeks a year. How my parents found the remote little caravans we sometimes booked, I don't know. Scottish readers must forgive me for pretending to know what Scotland is all about: clearly, I have just the unbalanced view of a southerner, a casual visitor, and I am well aware of that. But it means a lot.

In my early diaries, recorded mileages day after day reveal some surprisingly long distances, and just to get there entailed hundreds of miles during a day. My father, sometimes maybe with a bit of help from my brother, would drive vast distances, at first in his beloved Ford Consul. Once 516 miles (and, on the return trip, 516.2 miles), starting at 4.45 a.m. and getting there after 8 o'clock in the evening, to reach a point on the Cromarty Firth. It was a good base from which to explore the very far north and northwest. These old notebooks record the overhead passage of RAF and Royal Navy jets – roaring Buccaneers often the loudest – naval ships offshore, trawlers unloading their catch, distances to headlands and islands visible from outstanding viewpoints, heights of lighthouses, the names of MacBrayne ferries to the Western Isles. Always the weather. Often the midges.

It was invariably brilliant to be in the Highlands and to play around on Scotland's wonderful coasts. Old photographs confirm what I always thought: Scotland's landscapes are the equal of any, anywhere. But of course, this is from the point of view of someone who was then immersed in anything to do with the Highlands and Islands, especially a wonderful book, early in a special series, the Collins

'New Naturalist' *Natural History in the Highlands and Islands* by F. Fraser Darling and J. Morton Boyd. Although a revised edition (hence the addition of Morton Boyd) it retained the original cover design by Rosemary and Clifford Ellis, a striking breeding-plumage black-throated diver. What could make a better design than this naturally brilliant op-art pattern of grey, black and white? Later I soaked up Seton Gordon's great books, with so much more on local legend and history, and the meanings of the Gaelic names. William H. Murray's *Companion Guide to the West Highlands of Scotland* was also eagerly devoured, full of information in an entertaining, accessible and inspiring form. It was a chat with Gower's bird recorder, Harold Grenfell, in his house above the Mumbles, that later led to my discovery of John Prebble, whose books on Glen Coe and Culloden stirred the imaginative pot even more. And there was always Andy Stewart on the telly...

If the music of my early summers at the reservoir was the 1959-vintage *Cathy's Clown*, and in the winters came the Beatles, then despite the distant ring of bagpipes welcoming us as they played over the field at the end of the local Highland Games, the song of our early visits to the western Highlands was Ike and Tina Turner belting out *River Deep, Mountain High*. It remains an evocative tag to those superlative holidays, in the way that music often does.

A wet day in Glen Coe. If it didn't rain, we wouldn't have the waterfalls. The natural scene here is now somewhat spoiled by the 'H&S' addition of fences and steps.

The view from our first caravan: Loch Leven and the Pap of Glencoe, shielding the entrance to the wonderful Glen Coe valley. A 2020 watercolour based on a 1960s black-and-white photograph.

Many of the special birds in Scotland, however, are in and around the eastern or central Highlands, the Spey valley and the Cairngorms. My preference was for the west, not necessarily so high but somehow so ruggedly picturesque, sublime: but get on top of the Cairngorms and the vistas are unparalleled, the dizzying cliffs falling from high plateaux scarily impressive. Are the permanent snowfields of the 1960s still there? Yet from a distance they never had the same appeal for me as the spiky western highland peaks and troughs, the sudden appearance of an isolated peak rising from lower ground all around, or row after row of high peaks, each identified and named from the appropriate beautiful OS map.

We did have the odd holiday in the east but mostly stayed farther west and went east on one or two days, taking in the Loch Garten ospreys, Slavonian grebes, trying to find a capercaillie. I was never a climber, but we did go up a steeper side of Ben Nevis, rather than taking the regular path, to stand on top in warm sunshine, in shirtsleeves. More often, there would be a gale. Ben Nevis is a bit of a shapeless hump from a long distance, but on the top, boy is it *spectacular*.

Our first holidays were beside Loch Leven, near Kinlochleven in Argyll, where we first met the McAngus sisters who owned the caravan in 1964. In 2023 some strange urge drove me to look up 'Ena McAngus' and led me to her older sister, Catherine, and eventually to a long phone call: an amazing link to those long-ago schoolboy days.

Here's a random July day in the far north. It was raining and I was miserable: all year waiting for *this* and it was wet and dreary and hardly likely to be worth going out. Why does it *do* this? It was a typically selfish reaction; everyone was in the same boat and doing their best. I was apt to feel hard done by and to show it. Off we went in the car anyway. To soften the blow a bit I was put in the front seat, which I appreciate now.

First bird: a splendid osprey. It flew right over the road above the car and repeatedly tried to land on a nearby slope, but was chased off by gulls. I instantly perked up. It showed off its underwing pattern and the white head with a broad black band, and even hovered – brilliant. The day was improving but how could you beat this anyway? Ospreys were still rare and special back then, with probably fewer than 10 pairs in Scotland and none elsewhere in the UK.

Next, a new bird, complete with underlining: a great northern diver. Nearby were eiders, arctic terns, a black guillemot (always a favourite with the soft grey-black, big snow-white patches and, if you are lucky vivid red inside the mouth as it calls). The diver was on choppy water far out in a bay, but appeared to have a black head: so out came the new telescope (a Nickel Supra, one of the fancy new all-black makes with a variable magnification and focus operated by little knobs – no brass draw tubes here). The black head, striped neck patch and chequered back were obvious; the heavy bill looked more massive than a black-throated diver's, but I noted that this would be a difficult distinction in winter, at least with this individual. Perhaps I had in mind the chance of finding one on my Staffordshire reservoirs, one day.

Being unable to confirm a probable great skua was a bit of a disappointment. But soon we reached Smoo Cave on the north coast. There were 10 or 20 pairs of fulmars. I had been reading James Fisher's amazing, huge book on fulmars, and Smoo Cave was mentioned in his comprehensive inventory of sites. It was, again, a mixture of what *little* I knew, what I could see, and what I had read about. The fulmars were great: I wrote 'The bickering and greeting calls of these birds echoed all around the cave and the long gorge-like inlet at its front.' It is easy to imagine how James Fisher became besotted by fulmars and their chosen coastal cliffs and islands, and could later say that his book might easily have been much shorter. He died, too soon, in a car crash and was apparently partway through a book on the gannet – it would surely have been a great one.

So now we were getting on fine and the day was a thoroughly enjoyable one. As we travelled, I noted stonechats, wheatears, a buzzard, more eiders, then a red-throated diver on a breeding loch. There were small lochans later with more red-throats and a north-Scotland special, a greenshank. Greenshank: yes, a regular if never numerous estuary bird, a migrant inland, but in northern Scotland a rare nester, one of the most romantic waders of bog and peatland flows, its nest one of the hardest to find. I'd read all this, so I knew. But not from any personal experience.

Loch Loyal beneath the multi-peaked Ben Loyal trumped that with both red-throated and black-throated divers. Three species of diver in one day. Typically, the red-throat was heard first, its loud *kwuck* calls given constantly as it returned from a fishing trip on the sea. A flying diver has its long bill, head and neck extended, straight or slightly drooped depending on the species, its back humped, legs and feet trailing, and its wings are long and narrow, much more pointed than, say, a cormorant's. Loch Loyal is a large and impressive sheet of water, but red-throated divers more often nest beside a little lochan and so must feed at sea (black-throats tend more often to nest and feed at a much larger loch, although we did see small groups in semi-display on sheltered coastal bays). So now and then a red-throated diver returns to its lochan, flying high, hump-backed, calling loudly as it goes – so, as usual, keep your ears open as well as your eyes. Black-throats in breeding plumage are surely the most perfectly drawn design in the bird world. They were, and remain, such a privilege to see.

The day finished with golden plover, greenshank again, snipe, goosander, whinchats, more fulmars. There was nothing out of the ordinary, nothing special to write to the local bird recorder about, but for someone from far-inland midland England, the whole thing – moors, mountains, coastal cliffs, caves and sandy bays, divers and greenshanks, even ordinary wheatears – all this made for a happy day with the family and all was well with the world.

But a few days later there *was* a bird to write home about, to send in to the Rarities Committee, no less, not just the local recorder. I found a white-winged black tern, a magnificent bird in sparkling black-and-white breeding plumage. Fantastic: a good half hour at close range allowed plenty of notes. It was, as a friend said, a long way north for one of those...

That same day we visited Dunnet Head, the northernmost point on the mainland, where there were hundreds or thousands of puffins, guillemots, razorbills and kittiwakes and always the chance of a skua or two. The headland is rather square and falls abruptly into 400-foot cliffs, with rectangular blocks and horizontal ledges, some more sloping rubbly sections at the top where puffins gather, grass-covered buttresses low down and a low skirt of heavy, fallen boulders washed by the sea. A low, chunky white lighthouse is perched at the edge in a little stone-walled enclosure, above which a scattering of little chimneys sprout from hidden buildings. Across the sea we could look at the misty shapes of Orkney, the improbably tall, thin pile of the Old Man of Hoy. Looking left, all along the north coast of Scotland, towards Strathy Point where we would photograph Scots primroses in the turf and watch black guillemots dotted offshore. My notes concluded with 'I could spend days at such a place.' If only, if only.

Big seabird colonies were always a huge treat for me. Given the settings – often something like Old Red Sandstone cliffs with myriad ledges and exciting plants to find, too – they are often great places to visit, birds or no birds, but the noise and action in a seabird colony make for constant entertainment. Compared with nearby Dunnet, the fabulous cliffs and stacks of Duncansby Head, just east of John o'Groats, with their gaping chasms, swirling inky-black, blue and green sea and brilliant viewpoints for the seabirds, is a more varied, specially favoured spot. The last time I went, ready to show my wife how good it was, it was all obscured by thick fog.

There were decent heath and bog habitats close to home, on the sandstones and gravels and extensive commons of Cannock Chase, Gentleshaw and Chasewater, but still I revelled in the whole sensory experience of the tall heather, the boggy, sparkling, colourful sphagnum mosses, the sun and showers and vast views over open sea. The birds: fulmars! I'd read about them, I knew a bit about them, I knew they had spread from St Kilda not many decades back: I had a bit of background from those great books. They *meant* something to me. Good birds and good places so often come back to an emotional attachment somewhere. The more I look back, the more I realise that I had done some prior reading – and peering endlessly at my beloved OS 1-inch maps – that also allowed me to relate what I saw to what I knew. A kind of affirmation of my little bit of knowledge, my hopes and expectations, and then the real thing on the day. All built up to a very strong and lasting impression. Stick a windfarm offshore and I'd be in tears.

This is a problem with all these places you fall in love with and to which you form such an emotional attachment. If they remain unspoiled it is wonderful. If they are damaged, as is so often the case, it is so sad. Damage can be anything: even a car park, a few 'interpretation' signs, a paved pathway, a fence, a visitor centre in the wrong place... on a scale rising to virtual annihilation. Increasingly it seems that boxes are ticked with no real understanding of what it is all about. Damage a marsh and all you need do is add a SANG (a suitable alternative natural green space): this is what local authorities so often do. The SANG (we used to call them fields, or meadows, but now expect children to have fond memories of their sunny summer holidays playing in the SANG) – the SANG is a bit of grass with a cycle track, a dog walkers' area, probably somewhere to fly a kite (nice enough), somewhere to kick a ball about. Probably, the SANG is placed on what was a decent little wet meadow to start with, rather than on a bit of old, dry field that no-one would miss. So you lose the wet meadow, the orchids, the comfrey and agrimony, the butterflies and grasshoppers, the sedge warblers and stonechats and snipe, the grass snakes and frogs and toads and hedgehogs, and you get a bit of bare ground where even a magpie would have second thoughts. And a load of signs telling you what used to be there, and how good this new initiative is for biodiversity. Wildlife, as once was,

Coming back to Scotland

Glen Coe is unbeatably brilliant. The glen itself, of course, carries a weight of history. The Glen Coe massacre is hardly mentioned these days, but as an example of inter-clan rivalries and duplicitous behaviour on behalf of the government, it is a chilling tale, especially if told well by a writer such as John Prebble.

From the south as the average English visitor approaches it, the road swings beneath the unbeatable, furrowed cone of Buachaille Etive Mor and the extensive remote flows of Rannoch Moor. Views of Glen Coe as they open up, with its stunning 'three sisters,' are among the most memorable and meaningful of the Highlands for me, although on my most recent visit (2009) I couldn't help feeling the mountains looked a little smaller than they used to. A friend who became a geologist told me 'there was too much rock'. Hmmm. Towards the coast there is a quiet, sidelined little village, and there we used to hang over the wall of a hump-backed bridge to watch the magnificent salmon in the swirling river below.

To the south, Loch Tulla has a pine wood close by. You may think of a pinewood as a plantation, dense and dark (although sometimes beautifully green in the sunshine), uniform in age and structure, probably all planted in straight lines, not as a rule great for birds. This, though, is part of the great ancient Caledonian pinewood, something altogether different, thoroughly beautiful and very special for its wildlife. Scots pines range from tiny stems to huge 'granny' pines, tall and broad, silvered on the trunk, rich orange on the outer limbs. There are humps and hollows, clearances full of tall heather where you can disappear as you walk through, banks of bilberry. Such forests are to be treasured. The one beside Loch Tulla was brilliant but was then deer-fenced, and this may, I think, correspond with the loss of a great bird that we used to see there: the capercaillie. My friend and in some ways mentor Tony (A. R. M.) Blake tipped me off about them when I told him we were heading north: there was a male that had sometimes come out and attacked him when he had visited the wood. The hooked white beak of a caper makes it a potential foe not to be trifled with. But for all their size, capercaillies and head-height fences do not get on well: the great birds fly smack bang into them, usually fatally. But to begin with, the Loch Tulla fragment was open, spilling beautifully onto the moors and towards the loch shore. Here's a visit in the late 1960s:

> We walked in pouring rain in a hillside wood of mixed areas of Scots pine and birch. Signs of capercaillies everywhere – droppings, feathers, scrapes and obvious resting and dusting places. Shed feathers were mostly the long black, square-tipped tail feathers of the males, all waxy and shiny with little white marks towards the tip. The birds themselves were flushed from the birch areas only; a male, probable female and about five juveniles, more or less together, the juveniles seen on the ground [*I wonder now why these were 'juveniles', not females*]. Second male nearby flew up from about 20 yards, showing brown back, glossy blue-green tail, red wattle over eye etc.; a male (third?) seen later and several females in odd ones and twos.

On a return visit two days later, in much better weather, we found no capercaillies at all. Still, there were great compensations, including black-throated divers on the loch, a merlin and no fewer than six golden eagles. All around are marvellous, magnificent high hills, the ideal backdrop to these great birds.

> We walked five miles from the nearest road and left behind all signs of habitation: it was a really marvellous place to be. Not only was the panorama superb, but there were red deer scattered liberally over the hills and then – the eagles! Two were seen early on over Stob a'Choire Odhair. Then much later two over Meall nan Eun, flew to Stob a' Bhruaich Leith; two over Stob Gabhar were seen at the same time, as were two farther east over the same hill, presumably the original two from earlier on.
>
> Two performed some quite spectacular rolling and diving manoeuvres; the two farthest east flew in long, fast, straight, rock-steady glides with scarcely a wingbeat. One of the middle pair flew towards us and then crossed to another peak, with long glides interspersed with several rapid beats followed by closed-winged dives; it covered this at what must have been at least 60 mph. It put up 30 hooded crows, which flew up high and

> stayed there while the eagle, now clearly seen to be an immature, was dived at by a kestrel, but it took no notice of the smaller bird. Later on, this one and two others were seen several more times.

Those long and difficult Gaelic names added an extra bit of mystique, although I could never pronounce them. We saw quite a lot of eagles over the years, but this was clearly a much better than average eagle-watching session. Anyone who knows them well will have interpreted all these dives and rolls and long straight flights, but to me they were simply 'display', to be appreciated. In his book *Eagle Days*, published by Ian Langford in 2012, Stuart Rae clears up many misrepresentations and misunderstandings, and shows that long, direct, gliding flights are as much part of a 'display' as the dives and switchbacks. They are emphasising their mere presence. Eagles are astonishing things.

Family holidays in Scotland in the 1960s and 70s were in July and August, so golden eagle display was rather limited. It was on a winter visit to the Solway that I first saw an eagle properly displaying, one day early in March, at Murray's Monument:

> Hung in the wind with legs partially lowered, head drooped and wings angled. Later gave a display flight, consisting of long, shallow swoops, each followed by a curving climb on half-closed wings. The bird became nearly vertical, appeared to look from side to side, closed its wingtips in to its tail and swivelled over, to dive down in the typical broad 'teardrop' shape, into another swoop. This covered a good deal of ground with no wingbeats.

Other times in Scotland

One holiday was based over towards Aberdeen, near Banchory. It was great, if not quite so full of spectacular mountain scenery.

> An adult golden eagle was seen briefly, followed by an immature. The young bird was chased by two ravens, then joined by two adults; all three soared together, with occasional short dives. One adult swept down beyond a hill, with deep undulations, wings open on the upsweep, closed for the dives. At the same time a tiny speck far away and far higher also proved to be an eagle, probably over Balmoral! The line of sight was such that, for a brief period, the closer three and distant one were all in view in the telescope at once. Although I have seen six in the air at once, I think four in the field of view simultaneously might be quite unusual! Later, while watching a hen

harrier cross the valley, an eagle reappeared high over the valley, and the young one flew up again, swinging south, then north, low over the moors. Several times it swept upwards, closed its wingtips in to the tail, while its 'shoulders' were still held well out (creating a big 'teardrop' shape) and it performed several steep, twisting dives. This gave excellent views but the most unusual thing, for me, was that the bird seemed to be calling – I could hear a shrill yelping (Dad said yapping, like a little dog), *quip quip quip* or *quilp*. It was slightly reminiscent of the *pik pik* of an oystercatcher but less sharp and ringing. At first it was hard to be sure but the sound definitely moved along with the bird as it crossed over.

This eagle-watching was tremendous stuff. There was bright sunlight all the time, so colours were easy to see even against the sky. The young bird looked dark with broad yellowish-brown or golden bands across the upperwing, rounded white patches at the base of the inner primaries from above, white bands on the underwing and a pale tail (not so white as some I've seen) with a black band. The head looked quite pale and golden-yellow in some lights. Adults were browner with less marked pale areas on the wings, but golden-brown wing coverts often obvious [*faded feathers that bleach to a straw colour over the course of a year*].

As I've now said a few times, I prefer the northwest of Scotland. Sea, sky, beaches, moors, extraordinary peaks, lochs large and small. Water of gleaming silver, intense black, or purest azure. Even a word can bring back so much. Recently on TV there was a mention of Oldshoremore, a remote place with a long beach close to Cape Wrath. 'Been there,' I said. 'I wrote it down in my flower book: we found some sort of gentian.' I looked at my ancient *Collins Pocket Guide to Wild Flowers*, in the gentians. There, alongside Scottish gentian, Oldshoremore in tiny black print. My cousin John, when I told him about it, supplied a photograph of the occasion, the grand, sweeping bay, golden sands, backed by dunes. One word and it all came back.

Now, there is the 'North Coast 500', a route promoted around the far north on the same little single-track roads that used to be so attractive and enjoyable decades back. Apparently, they can be blocked solid, and stopping to look at a passing hen harrier, or scan for an eagle, is often a hazardous enterprise. It may help the local economy, which is clearly good. But do these visitors really appreciate what they are experiencing?

Did I just ask that? It is another terribly pompous, arrogant and selfish question; it implies an 'I can go but you can't' attitude. But do they, really? Can they

Like most large eagles, a golden eagle is pretty much 'dark brown', but it still has an attractive patina about it and, of course, is the equal of anything in the air, with a supreme quality of elegance and ease despite its size and potential for high speed. It is also able to soar to an immense height.

appreciate the peace and quiet and remoteness, when they are part of a convoy of impatient motorists? Do many care, really? I hope so. I'm encouraged by looking at my old maps and seeing so much open space, so much that surely cannot all be disturbed and spoiled. If Scotland becomes independent, I fervently hope these places with such irreplaceable qualities will be looked after. Will there be windfarms on every skyline? Cables over every hill? Of course we need renewable energy, but I've seen these things (California, Spain) and they are not good. One more huge dilemma for the future, which is rushing so fast to become the present.

***

Handa Island off the far northwest tip of Scotland used to be inhabited and had a parliament and a queen (the oldest surviving widow). There are still small remnants of ancient habitations, but after centuries of island life the people were finally defeated by the potato famine of the late 1840s. They had to leave in 1847, long before the people of the far more remote St Kilda, west of the Outer Isles. Handa was an RSPB nature reserve when I visited, but has now been taken over by the Scottish Wildlife Trust.

The island is a block of Torridonian sandstone, an exceptionally ancient deep red-grey rock coloured with lichens, washed by salt spray and decorated with salt-tolerant plants that grow from the most sheltered crevices. There are pale sandy beaches between low, sloping rocks where the island shelves inwards towards the mainland, and it is here that you land. Part has become detached in a huge, upright stack, scored with myriad ledges that are perfect for nesting seabirds such as guillemots and kittiwakes. The cliffs are sheer but the top mostly flat and level, like a giant stump cut off with a chainsaw.

Torridonian is named for Torridon, a little place not far from here, and Loch Torridon, with great peaks close by such as Liathach. These places featured in books I had read years before, stories by Joseph E. Chipperfield about eagles and deer and the people of the western Highlands and Inner Hebrides. Again, the idea of a visit to Handa brought together all this prior 'preparation': *very* little real knowledge, of course, but a general idea, and a mild excitement somewhere in the back of my mind that I was going to see the places I had enjoyed reading about. A rosy-spectacled view of past Highland life, no doubt; the 'romance of the clans' without much regard to the troubles, deficiencies and hardships of life for so many, then and now.

There was always a link between the present and something I had done or read about before, reinforcing the excitement of the day, adding a layer of interest and a bit of extra 'connection'. Does it, when you're a teenager, come through as a bit of ego? A bit of arrogance if you go on and on about it to other people? I know this and I know that... probably I was such an annoying little oik (okay, I still am). But it all seemed so good and interesting to me at the time, not the kind of thing you might expect from a typical teenager.

Handa was and still is, I hope, a remote place. The local boatman used to take people over just as and when they turned up at the little bay at Tarbet (where you might see a twite), to deposit them on the beach. The two that took us in different years both drowned in fishing accidents in later years. The first, Mr Alistair Munro, told us of days when the swell crashed against the rocks and solid white walls of spray rose right over the top of the 400-foot cliffs. Handa lies just south of Cape Wrath, the wonderfully named northwest corner of the mainland, and nothing much lies north and west until you get to the Faroe Islands and Greenland. There is plenty of space, a very long fetch, for a big swell to build up in a northwesterly wind. The sea can sparkle silver-blue, be smooth glaucous, turquoise-green, or grey and black and navy with white streaks and heaving, rolling wave tops, powerful and fearsome. I was once sent to a much larger island, Islay, to write about the RSPB reserve there, and I started with the arrival at the jetty, where the sea swirled deep and glossy black in the shadow of the rocks. Staff in Scotland complained that the sea must be blue under cloudless skies – but the editor, to her great credit, was on my side and the black stayed in.

It is odd how the sound of a little outboard motor on a single, small boat, put-putting across to Handa, can sound so evocative and pleasing, when the sound of, say, a jet-ski, *whoom-whoom-whooming* its way across the waves, can just ruin the peace and quiet for miles around.

Late July:

> Although virtually all the auks had left, very large numbers of kittiwakes and many fulmars gave us plenty to watch on the cliffs. One kittiwake was still crouched on a nest with its dead fledgling. Fulmars were feeding young by regurgitation; some young fulmars were panting in the heat.
>
> The real excitement came with the skuas. Several great skuas were approached to within 50 yards and flew closer overhead at times. Then two were seen on a rock and we approached carefully, hoping for photographs. Suddenly the birds flew up and one came at me about three feet high, swinging towards me in an arc, banking over, gaining speed and woosh! with a roar of wings shot past within a foot of my head – I ducked quickly! One bird was bolder than the other but sometimes both would come, even from different directions, almost at the same time. The attack was swift and powerful, a long, low period of acceleration then a very fast sweep at about chest height, at the last moment swerving upwards – really unnerving and hair-raising! At times up to five were around during these attacks.
>
> Their usual call was a low 'uh', but occasionally when flying higher they made a strange shrill whistle. Several times they performed a beautiful display flight, gliding with wings in a 'V', wavering with neck extended and drooped, all the time calling a more rolling version of the usual note.

An arctic skua also dived more steeply and spectacularly at my head. On Fetlar and the Caithness flows, I would later watch arctic skuas displaying with a more dynamic high soaring and diving flight, accompanied by loud, wild, nasal *ya-wo* calls. Skuas are always high on the list of desirable species on a seawatch.

***

Golden eagles populate my notes quite regularly on Scottish visits, become much less frequent later (with some in Greece, Spain or the Alps), and were absent for a time before a joyous reunion on Crete in 2017, where unexpectedly active, displaying birds in autumn reminded me of their sublime magnificence.

Ptarmigan appear in my notebooks intermittently: very occasionally, to be honest, and not for a long time. Will I ever see another? There were odd ones noted on isolated peaks here and there, but most were either on the Cairngorms or on a high western plateau that can be reached by the fantastic high, zigzag road to Applecross.

23 July 1968 was evidently a special day. For a start, at Loch Garten, three ospreys. But nearby, a new species: capercaillie (two females, seen rather briefly in flight). Later, near Loch Morlich, another new bird: long-eared owl. And in between, making a trio of new species for the day, two ptarmigan at 3,000 feet on Cairn Gorm:

> Watched at 25 yards range, mostly in mist but occasionally the cloud lifted to allow better views. Both very grey birds, rather olive-brown on the wings and mantle. Dark line through eye but apparently no red wattles. Beautiful markings visible during the clearer periods. Originally located when a bird disturbed a pebble, giving itself away by the rattle!

These were improved upon a few days later when another foray onto Cairn Gorm produced 11, approached to within about eight feet.

Ptarmigan.

The Applecross road is sensational. Here you can more or less walk into ptarmigan habitat, as the road goes so high. The Ford Consul, complete with excited family, was a heavy car with just three gears, first quite 'high' and with no syncromesh: double-declutching, or starting from a sudden unexpected standstill on a 1-in-5 slope to shuffle round a tiny hairpin, was quite an experience.

Another sample from my notes from 1968:

> I walked from the summit of the Applecross road to the northwestern end of Sgurr a' Chaorachain, where I sat on a rise just north of a tiny lochan. From here I could see across Coir nan Arr and over to Beinn Bhan, and along a nearer range of cliffs. A golden eagle appeared and flew towards me, then disappeared into a deep face in the cliff face. Soon it rose above the cliffs and turned round to face out over the corrie again, when a second eagle rose up towards it. The lower bird rolled onto its back and extended its legs, while the upper one also stretched its legs down until their feet almost touched. One dived vertically at the other and some fine aerobatics ensued. This was much less than a mile away (quite close in eagle terms) and I had excellent views. One swerved out, faced the low 'saddle' on the skyline and, once set on course, never moved a muscle afterwards. Head-on at eye-level it looked heavy, square-bodied but long-winged, and the eyes were visible from the front. With upturned wingtips the great bird flew slowly by, passing within 50 yards. Then I saw the second bird on the same course but slightly higher and it, too, came equally close on a dead-straight course, slightly descending to a sloping hillside, then over Beinn Bhan between 1,500 and 2,000 feet. It passed over some deer, which it seemed to dwarf, and swerved behind a rise, lost to view. One bird was chocolate-brown with golden crown and nape, golden-brown on the upperwing coverts, buffish near the base of the tail and at the base of the primaries beneath. The other was darker, sooty-brown, with golden head and pale, greyish-buff across the upperwing and tail base. As it crossed the hillside it looked a little unusual, very dark with the pale head and upperwing band striking. Both had quite blue bills with yellow cere, dark eyes and yellow legs.
>
> Later on, we saw both again and one stooped like a giant peregrine, very large-bodied with its wings curved back to the tail in a big, bulging teardrop.
>
> There were eight ptarmigan in a little valley, close to the summit (2,539 feet) and around 2,150–2,200 feet high. Like the eagles, they were watched in bright sunlight, ideal conditions, at ranges of 10–20 yards, sometimes

> closer. They gave various low, crooning notes and frequently a faint, rather high-pitched, fluty whistle.

I'm a little surprised I didn't try to say which eagle was male, which female. A pair together can often be differentiated according to size as the female will be larger, but overall the overlap is so large that trying to sex an isolated bird by size is next to impossible (or, in other words, ought not to be tried). Ptarmigan are 'white-winged' grouse, extremely delicate and beautifully sculpted, especially around the head and bill, with the most intricate plumage details. Like a few other mountain birds they face uncertainty as the climate warms: they can only go so much higher uphill, or farther north, before running out of land.

My notes earlier refer to golden eagles much less than a mile away (pretty good) but they were not always so. August 1973:

> Inverpolly: two golden eagles. First picked out using 7x binoculars as a steady, dark speck above the southern peak of Cul Mor, at a range [*measured from my beloved 1-inch OS map*] of 5 to 5¼ miles. I switched to the telescope and saw a bird fly up from the peak; later another lower down. Then both in view for 15 minutes, performing a long, steady glide interrupted by occasional circling, eventually going out of sight behind Stac Polly (3–3¼ miles away). Even at this range they looked big and majestic, their regular, unwavering flight and long, broad wings highly characteristic. They had covered roughly three miles at a height between 2,000 and 2,800 feet.

But just four days later, at a coastal site, I was watching eagles over a much lower slope and rated them some of the best I'd ever seen:

> Two hours of eagle-watching. First, an adult seen briefly, then two soared out and went off out of sight. Shortly afterwards one hovered above the slope like a gigantic kestrel, the first time I've seen an eagle do so. A second appeared and for a few incredible minutes they both soared around, frequently diving down in their characteristic 'teardrop' shape with wings curved back, the carpal joints pushed well forward and alula sticking out very prominently. They hung in the wind in bright sun, their golden heads dropped down, catching the light, the fore-edge of each wing seemingly ruffled and also very bright golden. Occasionally one would sweep down, wings half closed, sideslipping with huge, feathered legs dangled, bright orange-yellow feet striking, to pitch on a rock about 700 feet up the slope (so, 350–400 feet above me and perhaps 400–500 yards away). Telescope views were fantastic, the massive, bluish bill and yellow cere and gape so

For years a buzzard soaring over a suitably epic highland landscape was as good as it gets. Now I can see a buzzard every day in Dorset, if not in quite such an imposing setting.

clear; overall dark, rich brown with irregular pale spots or blotches [*a result of older feathers bleaching*]. One stood facing me with its wings drooped, away from its body, in a really 'heraldic' pose. Both finally soared away, chased by two ravens, which looked quite small! Some while later both reappeared, one in a long, low, slanting glide before it settled on a rock. It later simply lifted off this perch and gained height, rising without circling and with no wingbeats before angling back its wingtips and going away in a long, descending, very fast glide. Later on again another appeared and this was clearly a juvenile with prominent black-and-white tail and white wing patches. Two ravens mobbed it but it turned on them and dived at one repeatedly. Its wings seemed almost to move alternately, often having a 'ripple' effect from the base to the tip. An adult joined it before both went away and we saw no more of them that afternoon.

On yer bike

It is not often that a birdwatching trip of mine involves hiring a bicycle. This one was in 2009.

Coll was a bit of a dream. I had a long association with north and west Scotland but mostly through summer holidays – often, through great good fortune, two a year for me while I was at school and for a few years later. I read all I could, bought all the maps, learned about Glen Coe and Culloden. The western Highlands and all their history made a deep and lifelong impression. But the islands, save a few day visits to majestic Skye, a short stay on Islay and views across to the peaks of Rum, or Jura, remained largely unexplored, and I had never been to the island of Coll. When the opportunity arose, I thought at the very least it was a chance to see a corncrake, and to renew my acquaintance with some of the best landscapes (and seascapes) Britain has to offer.

Oban provides several interesting distractions and has fine views from its higher points, while you are waiting to set off. Visiting Coll necessitates a trip on Caledonian MacBrayne's smart ferry, and the crossing from Oban was fine enough. There was Mull, famous now for its white-tailed eagles, Ardnamurchan of fond memory, various island groups in all directions. It was a little bit rough, not too bad... But it was the return trip a few days later that really took me back so many years to those early holidays and views of the Hebrides. Sunshine was chasing showers around the Inner Hebrides, revealing in turn the length of Coll, distant Iona and Staffa, Mull, the extraordinary low, angular shapes of the Treshnish Isles, Rum, Eigg and Muck and the utterly beautiful Ardnamurchan Peninsula, complete with its memorable lighthouse. The islands, with all their colour drained out by the wet weather, looked wonderful, each its own shade of softest grey.

Coll's quay is overlooked by a great whale jawbone arch, like something out of Herman Melville's *Moby Dick* (brilliant book, but maybe could have stuck closer to the brilliant film). Standing by the whale jaw, between my two journeys, on a day of roaring westerly 50 mph gusts, I watched MacBrayne's *Clansman*, due to call in on its way to Tiree. It appeared more or less on time, great waves bursting right over the top of the vessel in explosions of white spume. It would have been exhilarating, but I was pleased to have had the smoother option. It was, though, an exciting sight and even the locals turned out to admire the skill of the crew turning and reversing the vessel in to the quay, and to see whether their returning children would make it back from school that day or have to go on to the next stop in Tiree. An anxious woman working at the quay held a portable anemometer aloft to record the wind speed.

Not far from the quay is a small hotel; six miles on is the RSPB nature reserve. To get there, I hired a bike from the Arinagour post office, accepting the chance you take on a Hebridean holiday – any Scottish holiday – that the weather might turn for the worse with little warning. Light drizzle intensified to driving rain and gales: I was suddenly cold, quickly drenched and by the time I got back had practically dissolved.

In Arinagour, tiny but more or less the village on Coll, I peered out across the bay, with its little stone piers piled high with colourful buoys and nets and lobster pots. Such places always have an irresistible appeal: they demand to be painted, but I never do it. A pair of eiders, the drake crisply black, white, cream, rose pink and pastel green, the duck intricately patterned in browns and black, played about close by while a merganser and some shags dived for fish. More unexpected was a marvellous great northern diver in breeding plumage, an incomparable chequerboard design. When I wrote an article for the RSPB's *Birds* magazine, referring to the diver, I consciously stole a line – 'Presently, I had something better to watch' – from W. H. Hudson, whose books I had mostly read, just as I sometimes pinched things from Dylan Thomas. Occasionally, a reader wrote to say it had been noticed, even appreciated!

Beyond the outer pier were the birds of the visit without a doubt: gannets, taking the 50 mph gales in their stride – or glide – as well as any albatross. They showed perfect poise, a genuine magnificence, brilliantly matched to the elements, taking the 'ordinary' gannet onto an altogether different plane. They headed so easily into the gale without effort, sharp wings slightly angled down and sweeping up to the tip to add a touch of extra refinement, while others wheeled and careened far offshore against the romantic backdrop of Mull and the Dutchman's Cap. Some came close enough to show the glint in their eyes, the sheen on their golden-ochre velvet heads. Birds heading so easily into the teeth of a gale are always fascinating: I've watched great black-backed gulls simply gliding fast along a small clifftop, head-on into a gale force wind, with no trouble at all.

It was on Coll that RSPB researchers investigated the precise needs of the disappearing corncrake, and how to supply them, paving the way for financial packages and advice for farmers. Corncrakes can't live with early cutting of grass for silage: an old-fashioned hay meadow was their ideal, but farmers get much more value from uniform grass grown fast and cut early, before corncrakes have had chance to breed. Coll has an abundance of little rocky outcrops, so, as the grass was mown or grazed, little islands of tall grass survived around the rocks. Coll offered corncrakes a lifeline that other islands such as Tiree did not.

Glorious gannet – revelling in 50 mph winds while the Caledonian MacBrayne ferry struggles: even the locals came out to watch as the ferry approached the quay.

The ride to the reserve took me to the warden, Ben Jones, the only RSPB staff member on the island. He told me all about corncrakes and habitat management, showed me where they might be watched, talked them up as you would expect, but couldn't show me one. In this weather, who could?

We went to Feall Bay, where the UK's only non-mainland colony of sand lizards survives and, in summer, frog orchids are, Ben said, 'as common as dirt'. Waves rolled in on the beach against the gale, blending from deep green to glaucous, translucent pale blue, their curling tops ripped off and hurled back out to sea in a fury of white foam and swirling, gleaming, rainbowed mist. It was magnificent, but a group of eiders and scattered shags took no obvious notice, riding the swell right to the beach, while a grey seal poked his nose out vertically like a big, floating bottle.

Somehow, snipe were drumming, redshanks were singing and lapwings displayed, making me wonder how it could have been worth their while in the raging wind and lashing rain. They could barely be heard, yet skylarks were also clearly pouring out their unsurpassed but inaudible songs high overhead.

Many fields would soon be brilliant for flowers but not so good for corncrakes. Taller, rougher vegetation, with a greater fertility, and more slugs and snails and insects, would be better. When the RSPB bought the reserve in 1991, the island had only 20 calling corncrakes: that rose to 180. In 2008, a year before I was there, there were 66 on the reserve alone, as well as 81 pairs of lapwings. Corncrake fortunes rose for several years but more recently have dipped, as the early enthusiasm for corncrake-friendly farming and support payments seems to have waned. As Ben told me back then, things are far from sorted. Take your eye off the ball and they will be gone again.

Twites are Hebridean specialities. I saw a handful, but less well than I had hoped. Twites interested me, as I used to see the regular wintering flock at my local patch in Staffordshire, before their disappearance as the habitat was changed and breeding numbers in England also plummeted. The bike ride also produced linnets – more than twites, prompting the usual thoughts about why do you need two such similar species when one might do? – and a cuckoo, which would be interested in the local abundance of meadow pipits. Groups of greylag geese were interesting, too, as they might be part of the original native breeding population rather than the introduced birds I was used to in southern England. They have become so intermingled, it is now hard to tell, but this seemed a good location for wild ones. In a lull in the wind, the song of a grasshopper warbler was a bit of a surprise and then, very briefly, came the rasp of a distant corncrake. But that was it.

The return ride to Arinagour was against reverberating gusts of wind in awful, stinging, freezing rain (was this really May?) and it was hard to see anything at all, or even to be bothered anyway. It was all too painful. Yet Coll, and the ferry trips from and to beautiful Oban of long-ago memory, had left its mark: a brilliant place.

GREAT SHRIKES

A bird that used to loom large in my notebooks is the great grey shrike: great bird, great name! Here are some old sketches of different postures. You can see the bold patterns and extravagant shapes, with the tail used for balance. Pale grey, black and white make a beautiful combination on several species, but it looks especially good on a great grey shrike.

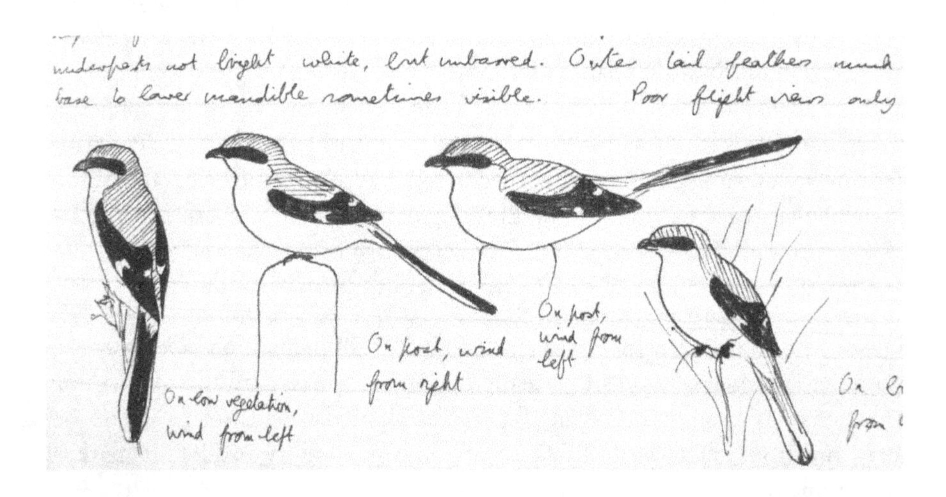

My first great grey shrike flew in front of me, a schoolboy, as I was walking along minding my own business on the Essex coast, after visiting my cousin's farm. It settled on top of a thick old hedge, so I could get a good look. Later, there was one, sometimes two, at my local reservoir for several winters and one, sometimes two, up on Cannock Chase, too, so I could see two or three in a day. I might have seen four but would have to search through years of notebooks to be sure. There would be an occasional one while I lived in Wales, either on Gower or maybe up at Tregaron where we went to see the kites. Years later, one appeared on the RSPB reserve in Sandy, not so far from the office... a bit different, as it preferred to perch high up on tall conifers overlooking an area of clear-fell. The New Forest has thrown up a few, too and one was so close that, despite all these previous encounters, I reckoned the views were my best ever, but that may have been largely because my optical equipment had improved a good deal over the years. The odd one will even sing a bit.

But shrikes of all kinds are good to see: difficult, though, in the UK. Great greys are rather scarce (not to say rare in most places) autumn and winter visitors. Long ago there were breeding red-backed shrikes: I managed to see a couple of the last pairs in East Anglia. Now they are rare migrants: you can't just expect to see one, unless you are very lucky, or hear about one and go to look (twitcher!).

Shrikes are smallish (thrush-sized) songbirds but equipped with sharp, hooked claws and slightly hooked beaks, and they act like little birds of prey. I've seen a woodchat shrike chase, catch and kill a whitethroat, for example, and great greys apparently take a lot of chaffinches (in the New Forest, maybe even the odd Dartford warbler). It is not easy to see, but they have the habit of impaling prey on thorns, both to help them deal with something far too large to swallow in one go, and to store surplus food for later.

Great grey shrikes are pale grey, white and black and, from a distance, one may look like a distinct white 'blob' on a treetop, bush or phone wire. Frequently they use a high, open, obvious perch with a wide view, so they are at the same time easy to see. Yet, many many times, they have proved unbelievably elusive. They either sit inside a bush, or perch low down, usually choosing the far side so you can't see them! This is one of those birds that can be around all winter, yet, on the day you take someone out to see it, the shrike will just be impossible to find.

An area of heath and bog, interesting for its flowers as much as anything (orchids, cranberry, sundews, butterwort and the like), was turned into a landfill by the admirable local authority. A great grey shrike, one of two present in the area, nevertheless took advantage. One January day I entered in my notes:

> One north of the pool as usual and one 1,500 yards away, east of the dam, as usual. The latter was seen especially well. Flew down to an area of grass and hedge cuttings on the tip and took an item of food in one minute; later perched on a bush 12 yards from the tip. Flew down to take food from the tip side (very quick), twice, just like normal feeding from the ground. Later, when it appeared to have left, after I had inspected the tip, I flushed it at very close range and it flew off to chase some house sparrows nearby on the tip edge; it soon returned, flashing just past me at close range in pursuit of a dunnock. Dashed about amongst willow bushes, then chased a dunnock back towards me; it hovered over the intended prey at no more than 15 yards from me, its bill half open, then settled at equally close range in a bush – exceptionally good views. Flew back to another bush and disappeared inside, later on an open bush before returning to the tip. Later perched on a tin at the rim of the tip, looking around for several minutes;

> then feeding on the flat top of the tip, which appeared quite lifeless (rough, stony earth etc.) although the shrike obviously took food several times. Either standing on flat ground or perched on small eminences, darting forward to take something from the ground, looking very wheatear-like. The tip is largely of flattened, burned tins and assorted refuse, covered over with grey shale and ash, with areas of bushy grass and hedge cuttings. Where the shrike initially fed it seemed quite 'dead'; in patches with rotting potato peelings etc. there was an abundance of small flies, like house flies – the shrike didn't seem to be catching them. Beneath the layers of cuttings could be heard dozens, or hundreds, of house crickets, but only by moving the grass away could I see these fat, brown, odd-looking insects. No doubt they would be relished by the shrike if it could get them.

Hmm, lovely... crawling about on refuse tips was not a normal occupation of mine. But watching shrikes was: this reminds me of the kind of close views that I was often getting.

> Great grey shrike found in an isolated rowan tree. It stayed in the same spot for a long time, then dropped to the ground, came up to a smaller bush, then dropped out of sight again and came back to the original tree. It moved from a near-treetop perch – obvious with the naked eye at 150 yards – to a perch within the small twigs halfway down, virtually out of sight.
>
> The intervening ground was all open bracken, so, instead of trying to get closer, I sat and watched for a long time – well over an hour – mostly with the telescope, getting superb views. It was a classic great grey – top bird, always a favourite – sometimes with the wind catching the back feathers, more often the lower throat, creating a slight raised ruff effect. This was enhanced when the bird began to call – and then to sing! I could see the throat moving and heard various harsh, low, slightly squeaky, slightly grating or rasping notes, like *chu-swee* or *tshwairk*, with various slightly more musical, short, bubbling, chirruping sounds added at times. I may have heard calls before (can't remember) but not song or even subsong so far as I can recall. After even eating lunch while still watching it, we left it still there in this isolated tree with its adjacent small bush.
>
> Its white superciliary was narrow and diffuse in shade, but in bright light (it was sunny throughout) the whole forehead and area above the mask was diffusely whitish, creating a pale, ill-defined surround to the mask. As it preened, I could see all the patterning well, including the underwing – white coverts except a grey band across the primary-covert area, whitish

secondaries with dark lengthwise streaks (a thicker, darker streak on the outer secondary/inner primary); primaries white, darkening quickly to blackish tips.

...

I didn't find the shrike for 15–20 minutes, on a dull, grey, cold day, but when I did it was slightly better than before in a way, as the atmosphere was still and clear: although it was quite distant, in the usual area. Then it flew and I lost it for a time, but suddenly picked it up again, in a bushy tree (crab apple I think) right by the road – I watched it for a long time from 30–35 yards. This must be the best view I've had of a great grey shrike – I might have had similarly good views back in the 60s/70s, but not with such a good telescope, and anyway, these must be the longest 'best views'! It caught a couple of bumblebees from the ground, one of which it pulled to bits and ate, the other it seemed just to drop into a small gorse bush and then ignored. Brilliant.

IF AT FIRST...

If you try to see rarities, you have to expect disappointments from time to time. They tend to fly off eventually. But who knows: persistence may pay off.

> 13 August 1983. The fact that I walked to the extreme end of Blakeney Point, on loose shingle, to see a royal or lesser crested tern, which had left just before I got there, didn't exactly brighten the day but it became a beautiful afternoon. It was fresh and windy, with big, bursting white breakers coming in from the deep blue sea, a vivid blue sky and most of what little cloud remained now being streaks of pale mauve-grey, against the palest eggshell blue. It was more like a good day on the west coast than the North Sea. A Kentish plover was bird of the day.

Royal and lesser crested terns are both large, pale terns with a long, bright orange bill, a really testing pair. I tried again a few days later:

> 23 August. The royal or lesser crested tern was being watched by others while I was a few hundred yards away, but once again had gone by the time I arrived. This real disappointment was made all the more difficult to accept because I had such splendid views of all the other terns, brilliant, fishing over the sea or forming mixed flocks on the beach. All the time there were terns around, anything between 30 and 500 on the beach, ready to be scrutinised. About 500 each of common and Sandwich, plus 10 or a dozen arctic skuas, 20 fulmars and 100 common seals.
>
> ...
>
> 27 August. A superb morning mostly in bright, warm sunshine, with Peter Hayman. 'The tern' was seen at last, not far offshore, but for only about ten minutes. It was a fine lesser crested tern and we were able to show many people Peter's measured scale drawings of Sandwich, royal and lesser crested terns for comparison. There was also a juvenile dotterel, which we almost 'trod on' at a few feet range – an exquisite, perfect, beautiful bird with enormous, liquid, black eyes... Arctic skuas were chasing the terns, which included a brilliant adult roseate tern. A wryneck rounded off the day.

This tern was a difficult one, and the 'if at first you don't succeed' mantra applies to lesser crested and royal terns in the UK as a whole. Some rarities have increased, others have maintained a steady few-per-year occurrence that seems, over time, to reduce them a little to lesser rarities, some seem to increase only as observer numbers rise: but there are those that remain 'extreme'. These two terns are firmly in that category.

Most people thought the Blakeney tern was a royal tern, mostly agreeing with well-known people who put a name to it early on. Some were clearly unhappy about this but, looking at the scanty literature and hearing comments from people with more experience abroad, understandably thought a lesser crested should look smaller and shorter-billed than a Sandwich tern. Peter Hayman's drawings based on museum specimens showed the real situation.

There was no way I could contribute to this debate, but I found that my own notes from The Gambia (where I watched lesser crested, royal, Sandwich and Caspian terns together) suggested lesser crested would be smaller-bodied than Sandwich head-on, but bulkier and deeper-bodied in a side view, broader-winged and with a fractionally deeper and longer bill. (I have to say that these African notes would have been a bit skimpy, as there was so much else to look at, at the time.) The Blakeney bird nevertheless accorded with these notes and Peter's drawings, but some people with far more experience and undoubted ability still leaned towards royal. Eventually, as more people saw it for much longer and the discussion continued, it was officially accepted as a lesser crested, one of several that hung around for a while in northwest Europe over the next few years, even pairing with Sandwich terns.

A royal tern at Kenfig in West Glamorgan in 1979 also took a long time and much research to become accepted as such, while a lesser crested on Anglesey in 1982 was being debated by the authorities just as the Norfolk bird appeared in 1983. There was a good deal of disagreement, but this all served to draw out contributions from people who knew the birds concerned and helped sort out the problem of identifying these large orange-billed terns. The Anglesey one became the first British lesser crested while the Kenfig royal was the fifth in Britain and Ireland. Matters have since become even more complicated as royal tern has been split into two species, American and African, and both are now on the British list.

If you find one, call for back-up. Dial 999.

OCTOBER RARITIES

Stripey warblers

Oh yes, rare birds are exciting, but it is so easy to get hooked into chasing after all and sundry and seeing very little. It is best, very often, just to go out watching birds in likely places and, who knows, you might find something. Some rarities remain very clear in the memory; a few seem to have been all but forgotten, giving a little shock of surprise when I look back through my notes. Obviously, that's not a reflection on them... just age. But certain birds are undoubtedly more valued than others for whatever, personal, reasons.

I used to live 100 miles from the Norfolk coast, so a weekend drive up there was not too bad, but nor was it just up the road – although I did once go before work, to see an ivory gull. We found the dead seal, on which it had been feeding, but did not see the gull. I've missed others, too.

Not long ago, nobody really knew much about what was around unless you could join in 'the grapevine' with usually weekly phone calls. It was a ten-minute walk to the phone box on the corner. Dial, feed in the coins, press button A:

'Mrs Phillips speaking.'

'Oh, hello, I was trying to get Eric.'

'He's out, but if you've got any information for him I can put it on the pad; if you want to know what's around and where it is, I can read out the messages to you.'

Most of Eric's intelligence came from Dave Holman in Norfolk and assorted regular contacts in the grapevine, and I gratefully accepted what was on offer, even if only rarely acting upon it, usually on a day out with Eric and John Fortey.

Then came Birdline, and you could just phone a number and get a recorded message, listing things that had been seen, and head off confident in having a good day on the morrow: but so often ending up seeing nothing. Friday departures, so many of them, not Birdline's fault.

Norfolk in October is one of *the* places for yellow-browed and Pallas's warblers. Yellow-browed used to be rare but in some years recently as many as 1,000 have turned up in the UK in a single year. That has not done anything to devalue them or reduce their inspirational effect. Pallas's used to be practically unknown; although also now appearing in larger numbers, they are still pretty rare. One day I was up there in Norfolk and had a good time, but no rare warblers appeared. The next day it transpired that two Pallas's warblers, two yellow-browed warblers, a little bunting and a Richard's pipit had appeared and at least one Pallas's had stayed all week. So, off we go again.

Arriving at 8.30 in the woods near Wells, I gave myself plenty of time. After all, to see my first Radde's warbler in the same wood had taken eight hours over two days. I was there until 5 p.m. In one place people told me that a Pallas's had been seen there the previous day. Then someone coming back from the end of the wood said there was a Pallas's there *now*, being watched as we speak! It had gone when I reached the spot. An hour and a half there proved fruitless, a waste of precious daylight. Ah, but more people arriving said that one, if not two, were being watched closely, by *crowds*, back at the other end of the wood... and another was found in a different place entirely. After failing to see this new arrival, I later heard that it reappeared just after I had left. So, I headed on towards the crowd-pleasers.

Halfway there, someone on the track was peering up into a tree.

'Anything?'

'Oh, it's a funny yellow-browed, with a great big crown stripe – a lot of people would confuse it with a Pallas's.'

At last, I saw one of the birds. After watching it as long as I could, I went off to the 'two Pallas's' spot and found both had gone. Top lister Ron Johns went into the wood, determined: later he came back and said he'd seen one. I gave up.

All was by no means lost, however, although this was so often the way with Norfolk: a long stretch of coast, a lot of people looking, a lot of reports, real or imaginary, a lot of chasing around seeing nothing. Sometimes there was no silver lining. But on this day, I did see that 'yellow-browed with a great big crown stripe.' It was described as such by someone who had already seen two Pallas's that morning. At first it called softly from a belt of small sallows. From them came quiet *chip*, *cheep* and more nearly disyllabic *chweep* notes, none quite like the usual calls of a yellow-browed (variously like a thin, piercing *ts'wee* or a more distinct *tchi-weee!*, sometimes repeated quickly for a few minutes). As soon as I saw it, it seemed that the head was remarkably well contrasted – and was there even a glimpse of a pale rump? (Both have strongly marked heads with bold pale superciliaries, but

Pallas's has a clearer central crown stripe edged more boldly with green-black than any yellow-browed, as well as a pale lemon-yellow rump patch. Both have two striking pale bars on a dark wing.) The wings remained firmly closed, the rump unusually well hidden underneath. Then it leaped out, turned head-in to the bush and hovered: yellow rump! Wowww, Pallas's without a doubt.

A year later, another day, another list of birds not seen. Six or eight Pallas's, a yellow-browed, one or two Radde's warblers, a firecrest and a red-breasted flycatcher. My score was poor, with two or three Pallas's, but as I noted, even one would potentially be bird of the year.

After four hours of listening to other people saying what great views they had had of nearly all these things, I had seen nothing. Then a flash of a stripey head, a yellow wing-bar – couldn't be anything else! Pallas's warbler. It was quite some time later that I got it in my sights properly and watched it well for some minutes. Plenty of notes, plenty of activity with this very active little bird.

Then it or another (was it the first, second or a third?) gave great views, being watched by a little knot of people, and after that I found one for myself. One looked dull and greyish, this latest bright and green, so I reckoned I had seen certainly two, probably three. Real satisfaction, despite all the other rarities being missed.

There were many such September/October days in Norfolk, with frustration and elation in equal measure, yet such was my delight in these two double-wing-barred warblers, that a day with a good view of a yellow-browed or Pallas's would always be a happy one, whatever else I might have missed. My expectations were relatively low and my pleasure at even a moderate level of success, compared with most of the big listers, was sufficient. Sometimes there would be two or three yellow-browed warblers hanging around a big willow in a conifer plantation, or perhaps a patch of tall sycamores on the edge of the footpath, and it was just wonderful to be able to watch them – all the way from Siberia after all – and soak up the experience. Firecrests were scarce in Norfolk and now and then one would add to the day, and it was difficult to say, in truth, whether a firecrest looked better than a yellow-browed or not... probably, but the yellow-browed always had that extra charisma about it. They remain great favourites despite so much more experience, now, of so many fabulous birds in many countries.

I have in recent years seen the odd one on the south coast. They can be difficult.

> At last! My slight (!) obsession with yellow-browed warblers expressed by another visit: this time, I searched around for half an hour or more, then

> heard the bird call, then hung around for maybe three hours before I finally saw it. It eventually gave me superb close views.
>
> Called many times: sometimes two or three calls before a long silence, sometimes many more calls, but silent for maybe 15–20 minutes afterwards. Once an extraordinary series of almost continuous calls for perhaps two minutes (I heard this from an unseen bird at Studland a year or two ago) – this was a sequence of *chi-wee* or *tsiwee* notes with short, harder notes interspersed, like *tsiwee-tswi-wee-tsiwee-top, tswiwee-tsiwee-top-top, tswieee*... almost as if singing (but actually not really like the song pattern). But even though it did this from just a few yards away, it was always somewhere hidden in dense, tall thickets of willows, alder, birch, ivy and evergreen oak, and I just could not see it. It called from an area about 100 yards long, an area where I have seen yellow-browed before.
>
> Eventually, however, the call came louder and closer than ever and I saw it in the middle of a willow with bamboo etc. growing through it – it fed, moved around a little, and preened. After a few minutes I lost it but then I located it again, very close, almost overhead – brilliant.
>
> It was a relatively greyish-green bird, more grey-green and cream than green and yellow, almost no yellow in it at all, but nevertheless immaculate and beautifully marked, a typically lovely little bird.

October is just a great month for rarities. My notebooks are full of notes and sketches. In October 1996 I saw a Pallas's at a different place for a change, on a rare visit to Landguard Point in Suffolk.

> Bliss! Two all-time favourite species and great views of both. Glorious, pristine adult Mediterranean gulls, in bright sun, at 30 yards – really, some of the very best views I've ever had, with every feather distinct! Against a dark, grey-brown sea they looked simply magnificent, stunning birds. And a Pallas's warbler... I'll never get Pallas's out of my system. I saw this one very well and still want more. All day it was elusive, giving brief, intermittent views in a patch of dense tamarisk with elder and lilac, in sunshine but a strong wind. Good views, though, with the telescope, of a very bright and beautiful Pallas's, but with long gaps in between – how could it disappear? I left, went to Felixstowe, bought something for my tea, decided to go home but then thought, no, why not have one more look – and suddenly it was calm, if dull, and there was a fantastic bird, suddenly easy to see. All its features were seen as it moved with frequent sideslips and dives through foliage, equally adept at rising a few feet; frequent very brief hovers. Flight

> stronger than the flitting, dipping action of goldcrest, more direct, slightly dashing, across open spaces. Even then it was able to dive into a five-foot-high bush and disappear for long spells. No calls heard.

A couple of wintering Pallas's in recent years have given excellent views, too, and always offer that real sense of achievement. What about a Pallas's in March?

> What a difference from old days searching for Pallas's in Norfolk, cold, long, hard days searching sometimes for nothing – this was a fantastic day, huge Constable skies, some great downpours of rain and hail, ice swimming across the windscreen as we went towards Dorchester, but also sunshine across beautiful Dorset, vistas to the sea – and the Pallas's warbler in view as soon as I arrived, in bright, warm sunshine!
>
> A great bird – Pallas's are always great birds, but this was often very close, in full view, in bright light, flitting about, hovering, showing off brilliantly. It made a goldcrest look sluggish; much more 'bouncy' than the average yellow-browed.
>
> It was, I think, as good as any Pallas's I have ever seen, really clear and close (under 10 yards), low down in thickets, mostly of bramble. It was a particularly yellow bird – at first, I thought the head was washed yellowish overall and the crown stripe was weak, the sides of the crown yellow-green, the eye-stripes green rather than blackish – but as I watched and looked harder, the crown stripe was clear and very sharply defined, especially in a rear view, between dark moss-green bands; the front of the crown stripe and the superciliaries were bright yellow, the rear superciliary long and pale yellow; the eye-stripe dark greenish. The upperparts were typical, sometimes looking a little greyish around the nape; back bright moss-green, upper wing-bar short and thin, main wing-bar (greater coverts) wide and yellow; tertial fringes whitish (yellowish?), tipped whiter. Underside almost all white, with some dusky marks on the upper flank just under the closed wing. Broad pale yellow rump band very obvious and frequently seen – often flirted outside a bush, hovered, or flicked upwards or downwards revealing the rump. Legs dark greyish with bright, pale orange-yellow feet. A superb bird.

Yellow-browed warblers in Dorset:

> Despite dire forecasts and heavy rain overnight/early on, I went back for another dose of Pallas's warbler disappointment: and it was a really beautiful day (only wet again after I left). No sign of the Pallas's for anyone

today and despite the overnight weather there seemed to be fewer birds (goldcrests) in the wood than before.

I heard, though, of a yellow-browed warbler on the edge of the marsh and found a few people looking for it. It soon called but at the same time I heard a longer burst of calls behind us – surely two? We saw the first and a second small bird quickly flew – not long afterwards there were two yellow-browed warblers about a foot apart! Then followed frequent views, from brief glimpses to very good, and the two separated – it seemed that two were calling to one side while the original bird kept to its bush, then we saw two again while another had been seen or heard 50 yards away. I think I probably saw all three at some time. Later, I returned and found everyone else had gone, so I had one bird entirely to myself and got even better views – initial views were in bright sunlight but the later ones were in softer light which enhanced some features, such as the very soft streaking on the breast. It kept largely to one big willow bush with a dense growth of bramble and bindweed underneath.

All the usual features – very subtle, soft paler central crown, small, short upper wing-bar on one, seemingly stronger on another; pale greyish legs and brighter yellow-orange feet etc. The dark areas on the wing – both sides of the main wing-bar and the dark tertials – looked darker and blacker from the rear 'against the grain', but soft olive green from the side. Beautifully marked birds with pale tertial tips especially striking at times, but the wing-bar and superciliary catch the eye about equally – wing-bar a really striking and obvious instant feature.

Calls strong close up, weak and wispy at any distance – sharp, bright, high *cheweee* or *tsooip* or *tsooee*, or *tseeoo*, strong start, thinner finish. Always to my mind a clear disyllable but sometimes more slurred into almost one.

Common stuff can still be appreciated. A week later:

House sparrows – 15+ in the garden. They are around the village but a bit erratic in and around our garden, but recently several have been on or around our wall and in the garden next door, sometimes coming into our garden to feed – good to see.

Yes, I still like my sparrows.

In the 1970s I visited the Isles of Scilly several times: here there was an abundance of rare birds, provoking an equal abundance of discussions and arguments.

Here's an October juvenile long-billed dowitcher from North America (my first), seen after a long journey, but well worth the effort. None of us could really afford the share of the petrol. It was the first of several dowitchers: one of the more recent ones, I found for myself just 10 minutes from home. It would have been so much cheaper to have waited.

A Richard's pipit was thoroughly discussed over the wall as we watched it on a field on St Agnes: was it perhaps a Blyth's? I had no experience to draw on and said nothing, struck dumb before noted star observers, but although it was pale and had some strange, soft calls, it was, I think, a Richard's after all.

On another occasion, a rose-breasted grosbeak appeared in front of several of us stunned watchers as we waited for something else to appear: so it was a kind of 'multiple observer' find. It was the sixth record for Britain. 'What a grimmler,' said Peter Grant in the pub later; 'what on earth did you think it was?'

An American robin was another great bird on beautiful St Agnes: as we heard about it and set off towards the Great Pool, I remember local Isles of Scilly expert, David Hunt, saying to everyone 'Don't run! Don't scare it away!' It wasn't scared away and remained for several days.

A Richard's pipit after all...

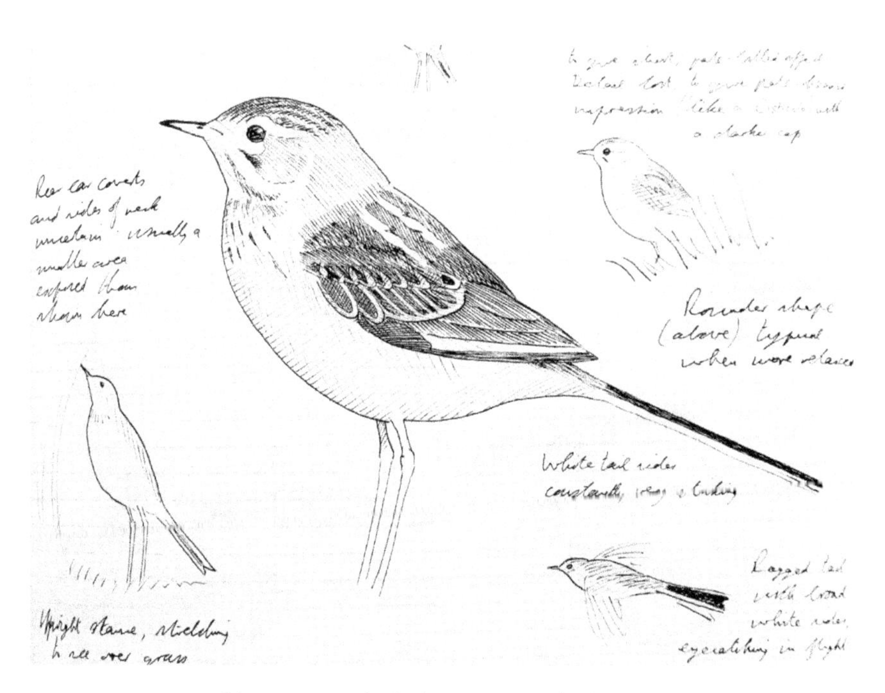

This one's a real Blyth's pipit, watched in Kent.

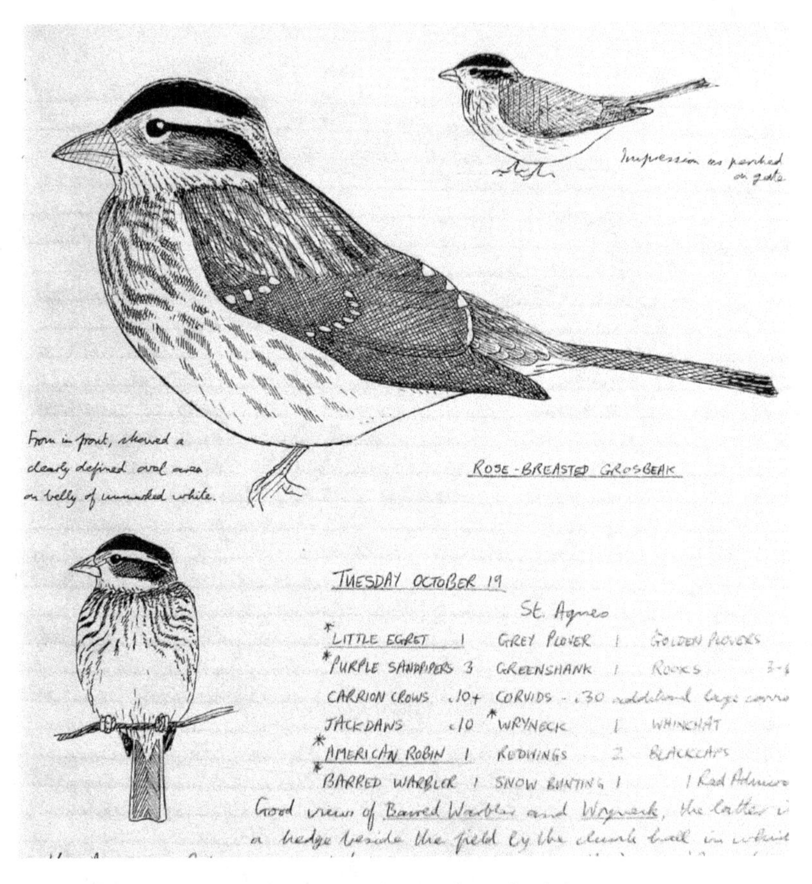

A bonus rose-breasted grosbeak.

The American robin stayed on St Agnes for several days.

The two-bird theory

This is very useful sometimes; very annoying at others; very suspicious, as a rule, especially if the first is a 'Friday bird' that has gone on the Saturday.

Say a white-winged black tern (rare) is reported... next day, a bunch of people arrive wanting desperately to see it. There it is, thank you, tick. Later several more people arrive: 'Oh yes, it's still here, saw it just now.' The second group see a black tern (regular). Ah, well, the white-winged black must have gone. And a black tern just turns up?

Now, was there really a white-winged black tern the previous day? The veracity and ability of the original observer might be questioned without any evidence either way. Unfair. If there was, had it been replaced by a black tern this morning? Two birds? Is the first group just a bunch of stringers (seeing a black tern and claiming it as the rare white-winged black)? Or was the white-winged black there earlier, or might even still be there somewhere, but a black tern has since appeared to complicate matters? One bird, or two? A reliable observer, or not? Reputations can be ruined (or made) in these situations.

It has happened to me a few times. A semipalmated sandpiper was at Keyhaven on the Solent. It was brilliant and gave me excellent views when I went to see it (I didn't find it), sometimes alongside a little stint for comparison. We all saw it. I took loads of notes. A well-known twitcher arrived. 'It's over there in the far corner of the pool, with a little stint,' I told him. There they were, two birds, side by side: two little stints.

The semi-p came back, fortunately (not that it mattered much to me: it had been seen well by plenty of people). It could have been a little awkward, though. Even then, there was a vague suspicion amongst the newcomers that we had seen a little stint while the semipalmated sandpiper was temporarily away – 'you've been watching the wrong bird.' No, we hadn't. But all kinds of arguments can ensue. The two-bird theory.

One year I went to Ireland, mainly as a trip to find my wife's long-lost relations near Cork. There was a flock of small waders by an estuary at Clonakilty and I looked through carefully. Wow! A Baird's sandpiper! A pretty good rarity, an excellent find. Two days later I took another look. The wader flock took flight as I arrived, as a peregrine flew over: a flicker of a white rump, must be a curlew sandpiper. Later they settled again: no curlew sandpiper. Odd. No, it's a white-rumped sandpiper! Another American wader. Did I make a mistake? No, the

Two American sandpipers in County Cork, 1994: White-rumped and Baird's.

first was a Baird's sandpiper, no doubt at all, but now here was a white-rumped instead. Two excellent finds. A genuine two-bird moment...

Feeling superior?

There is a risk, which I try hard to avoid but which sometimes inevitably comes through even in a book like this, of unwittingly feeling superior in some way to less experienced birdwatchers, in certain circumstances. Unwittingly revealing it. Wittingly, at times. Try not to say anything. Experience doesn't necessarily equal expertise, or vice versa, so you might be talking to a genius teenager or someone who has watched birds for 50 years but still can't do warblers. Or someone who has just a very superficial interest but has all the gear. It comes out as frustration, irritation or annoyance, sometimes, but shouldn't. It is the same with any similar specialist interest and may be followed by a mild or stronger feeling of guilt at having succumbed again to this unbidden reaction.

It might be, say, at a Silverstone Classic meeting where you can wander around and look closely, even hands-on, at wonderful old racing cars lined up in the paddock. Look at that, the engine's running and you could balance a glass of wine on the radiator and it wouldn't even shimmer. Just *look*! That, to other people, may be just another old red car, heavy and slow, a bit poor really, let's move on to something better.

'*What*?! That's *a Maserati 250F* for goodness sake. One of the most beautiful racing cars ever devised: the kind of car that Fangio and Stirling Moss drove, the kind of car they would drive at close to 200 mph wearing a little helmet and goggles and a short-sleeved cotton shirt, no seat belts... what do you mean, *just another red racing car*?'

Or it might have been at an air show in the later stages of the last Avro Vulcan's display career, as the great plane swept by – 'Oh, it's a big old plane, heavy, slow. Some sort of bomber I think.'

'What? I've known Vulcans *all my life*: I've watched them at Farnborough airshows – you know, Farnborough, where the first prototype Vulcan was rolled in front of the crowd; *rolled*! I've seen them at RAF aerodromes – the Vulcan, immense, unique, one of the V-bomber force, the nuclear deterrent, the one you'd scramble in the event of a four-minute warning, the one that bombed the Falklands – my stomach has rumbled to the sound of its four fantastic Rolls Royce Olympus engines, thrilled to the howl of the air in those huge intakes – it's the Vulcan, *you will never see it's like again*. You know, it wasn't always camouflaged like this, it used to be all-white with a bit of anti-dazzle black on the nose... I mean, you might not know, but *I do* and...'

'Wait a bit and there'll be something better come along I expect.'

But it *means* so much...

It might be in a hide in winter, looking out across a bit of reedbed, and someone peers out and sees 'Oh, a coot, and there's, what, oh yes, a mallard I think, and a Canada goose, and there's that bittern people have been seeing, and a cormorant and, what, that must be nearly 30 birds we've seen this morning!'

For goodness sake, I've spent *decades* without seeing a bittern as well as this... look, a bittern, it's a... a... it's a *bittern*! Look at it! Give it more than five seconds anyway. *Appreciate* it!

It comes through the experience of many years knowing and understanding that a bittern is not a bird you will see well every day. Or even every year. Experience in this sense means from books, or other people: you don't go around thinking 'I'm not seeing many bitterns this year'. Where do you learn these things? Not from the internet. Now, I'm all for – and always encourage – people looking at coots and mallards and thinking they are great birds (they are, after all), but just for a moment it might be as well to linger on that bittern, you know, it might... oh, it's gone. You might not see another. Bitterns used to be incredibly rare in Britain, you know – a pair or two at Minsmere and that was that. You'd never *see* one. You don't know how lucky you are...

There it is: *you don't know how lucky you are.* Talking (or thinking – I wouldn't actually say it) down to someone, patronising, expecting others to react as I would myself. It is an easy trap to fall into. Why not enjoy the mallard, if you prefer. But perhaps you might look up bittern when you get home – I mean, in a book, if you have one – and see what it means, how lucky you are, to see one like that.

Bernard Venables, famous decades ago for his best-selling *Mr Crabtree Goes Fishing*, wrote that a fisherman should really start young, to experience that first wriggling little perch on the end of a line... something that would remain in the memory for life. A little bit of magic. Build up from the common, easy little perch and you might one day be fishing for brown trout, or grayling, or specimen barbel. Start right in at that high level, catching a massive carp first time out, and somehow the magic is lost. You might feel you've done it all and give up. It reminded me of Nick Hornby's book *Fever Pitch*: he had watched Arsenal home and away for donkeys' years, and then along comes his girlfriend and they win the championship, first season she was watching. It hardly seemed fair. But to have had all those years behind him, to win the championship at last, was so special: he had the experience to 'appreciate' it, feeling he deserved it after so many disappointments, just as I think birdwatchers get more out of their interest as the years roll by. No reason at all why you shouldn't see a white-rumped sandpiper or a dusky warbler right off, but it seems better, somehow, to have had a few years to learn just how good these rare birds are, or at least to realise it. Not that I'm jealous.

Rare birds used to be just that: when I started, people would give out self-addressed postcards so others would send them information about rarities turning up. I used to do a weekly walk to the phone box on the corner to get the news before the weekend (we had no phone at home). Only later came Birdline, the phone number you could ring for info; and later still the pagers and real-time messages on your smart phone. But it was not just that. Radde's and dusky warblers used to be neck-and-neck with their numbers of records in Britain, around 30 or so slowly increasing year by year: now they seem 'lesser' rarities because so many have been seen, duskies anyway into the hundreds. To me, a yellow-browed warbler was a real rarity, a Pallas's warbler almost impossibly rare, hardly to be hoped for. After all, even the famous (or infamous) D. I. M. Wallace apparently hadn't seen a Pallas's warbler... Ian was already a well-known figure on the rarities scene and had a special liking for warblers (he had, after all, worked out how to identify the difficult *Hippolais* group, such as icterine and melodious warblers – most of which aren't even called *Hippolais* any more), but at that time I had never met him. It was not until around the year that I first went to the Isles of Scilly in the 1970s that Pallas's warblers staged a bit of an invasion and many of us saw one – or two. I remember the first on St Mary's, when someone said 'Oh, my heart!' (repeated years later by someone alongside me seeing his first wallcreeper). Now, Pallas's is a much rarer bird than yellow-browed, for sure, but still just 'scarce'. I still think of it as rare, and an exceedingly good rarity to see, too. I think I'm lucky in that way.

Great rarities that you would never expect to see, let alone find, have all disappeared from the 'rarities list', the birds that needed to be assessed by the *British Birds* Rarities Committee (BBRC). White-winged black tern was something to hope for, dream about, one you had to 'send in' to get a record accepted: now, barely a scarce migrant. But it is still the very same good bird to find. When I was chairman of the BBRC we had to reduce the numbers of species that were considered: some of the regulars with 20 or so records every year had to go. They could, after all, be considered perfectly well by local records committees, who'd know the birds better than the likes of me. But I tried to fight against it a little bit: it seemed wrong, somehow, for the annual Rarities Report to come out without Pallas's warbler, white-winged black tern and the like included. All those years of gathering records and the continuity was broken (periodic 'scarce migrants' reports helped put that right). Still, if I could find a Pallas's warbler next October, I would be over the moon.

I'm pleased I had my 'little wriggling perch' moments and didn't (or couldn't) leap into pagers and websites and endless twitching expeditions. My 'British list' remains small compared with so very many people now: kids in their early teens beat mine easily. Am I superior? Not at all, not in *any* way: nor do I feel that I am. It isn't that. But I like to have my memories, a little bit of nostalgia, a feeling that maybe I did things the right way round.

Blashford Lakes on the very edge of Hampshire has a hide overlooking a small marshy patch with clear channels carefully cut to give a good view through the dense vegetation.

> Brilliant bitterns: one difficult to see, but then wing stretching, and flapping about as it moved through reedmace, before coming into the open, fishing – superb. I have seen pictures of them, labelled 'drinking', with the head and neck arched forward, very low, twisted, with the bill almost sideways lying on or just into the water, but hadn't realised this was the pose used for long spells while searching for fish. Then I saw a second and had shorter but even better views, being closer: it was often gulping with a slightly open bill, pulsing throat and flicking tongue. The first was very sandy-golden on the neck and back, the wings greyer; the second looked paler, more buff and greyish. Superb.
>
> ...
>
> Far and away the longest views of a bittern I've ever had, probably in view nearly continuously over 40–50 minutes or so, perhaps more – brilliant. It caught two fish of 4–5 inches long and spent much time 'fishing' in the odd way, leaning its head and bill far forward, tilting its head over and dipping the bill tip into the water surface as if detecting vibrations or something, as much as looking.
>
> ...
>
> Bitterns already in view when I arrived (for once), then two gave easily the best ever views, beating those of a week ago – one or both in view nearly continuously, one coming as close as 8–10 yards. One was aggressive towards a moorhen, chasing it, and raising its crown and cheek feathers in a wide ruff. Someone suggested that the odd feeding pattern – used all the time, with the bill tip held sideways in the water, even thrust into dense vegetation – was 'using the tongue to attract fish', but I looked very hard and could see no evidence of that – in fact, even when a bird held its bill open several times, I could see no trace of the tongue anyway on this occasion. The long tertials normally just exactly cover the blackest, unmarked part of the wingtip, the rest (normally exposed) being finely marked. A third bittern flew by, the first I have seen in flight here.

My first bittern was at Oxwich on Gower. Peter Garvey and I tried to get to Oxwich Point, which looked interesting. We scrambled painfully through gorse and brambles, through hawthorn hedges and across wet ditches, over stone walls and up and down rocky slopes. We were chased by two obnoxious and ferocious dogs and told off by an aggressive farmer. There was nothing to see and, feeling despondent, we turned to walk back by a longer, easier route. We later bumped into a friend who had a van and offered us a lift back, but he needed half an hour

When I was drawing this bittern, at the end of winter, the lawn outside was vivid emerald in the sunshine, with a great show of crocuses and early daffodils, and there was a male blackbird, as jet black and perfect as could be. Why would a large bird such as a bittern be so brilliantly camouflaged (for defence against predators such as marsh harriers and white-tailed eagles?) while a similar species stands out in the open, bold as brass (grey heron, say, let alone an egret) and something like a blackbird, which really is vulnerable, stands out in such perfect, eye-catching contrast?

to load a stack of wood… we could have helped, but chose to look at the marsh instead. A strange, long-winged, pale sandy-brown bird rose from the reeds, legs dangling, a pale band across the wing. A bittern. We appreciated it.

It is easily possible – I've done it several times – to be looking straight at a bittern without realising it is there. When people see a bittern in the open, they may think 'Look at the amazing camouflage!' But to appreciate it properly you need to see a bittern in the reeds, tangled waterside grasses, reedmace and assorted muddy leaf litter that afford it such cover. It can be remarkably difficult to see it at all. Best if you lock on to its eye and bill, by chance. If it moves, head low, body horizontal, rising up and down with each step, it can look almost like a big cat slipping through the vegetation.

We (nearly) all make mistakes

Most of the scarce birds I have come across stayed to be seen by other people, or were co-found with others at the time, so the controversial single-observer question rarely arose. Sometimes, though, I have a little spasm of self-doubt, and begin to wonder...

My first rare bird should have been the most obvious, but I was still at school, rare birds were hardly even on the radar, and who would expect a hoopoe to fly by in November anyway? I blurted out something about 'a woodpecker thing', but that was not so much a misidentification as a way to get my companion onto the bird before it disappeared. Not a mistake, really, I tell myself.

Others have been momentarily more problematical, but usually quickly discovered and quietly put right, no harm done. We all see a bird flick into a bush, think 'blackcap' and find it is actually a garden warbler or something – not mistakes, exactly. Worse if you hope for a wryneck and it turns out to be a song thrush. Many years ago, I once got into a terrible mess with a supposed gull-billed tern that morphed into a juvenile Sandwich tern, and the resulting submission of the record was, I soon realised, a total embarrassment – even now I am not quite sure whether the original bird, before the Sandwich terns overwhelmed it, was real or not. And how did I ever try to fudge a dodgy dunlin into a broad-billed sandpiper, even getting other (more experienced) people to agree? It took a young Keith Vinicombe to point out our mistake. It may or may not have got serious – we never got close to submitting it – but it was a bad thing to do. Keith, perhaps, is an exception that proves the rule, the one who does not make mistakes.

I remember, too, on St Agnes, Scilly, one day in the 1970s, someone (Jeff Hazell) looking over my shoulder at a sketch in my notebook and saying: 'Red-necked? Are you sure? It's a grey, surely.' It was a phalarope I had watched closely at Chasewater a few days before, and I had not only convinced myself but also tried to convince the people with me that it was a red-necked. But, I suppose, it was a juvenile grey, still heavily marked with black-brown above. Bad mistake: there will have been others.

Years ago, people who discovered, identified and wrote about improbably rare birds on far-flung islands were in a different league entirely from us young birdwatchers far inland – they still are, come to that. One – although I did not know him, myself, then – who seemed to be thought infallible, a kind of gold standard for identification expertise and unquestioned reliability, was Fair Isle warden Roy Dennis. I don't know whether Roy felt infallible himself or not, or still does – I'll ask next time we meet. But I always remember his write-up of Britain's first Cretzschmar's bunting, and the words 'I flushed a roosting bunting from a field of rye-grass and, as it flew to land on a stone dyke, I gained the impression that the "jizz" was wrong for an ortolan bunting' made me realise how different things could be: just how good some people were. 'In flight, it appeared slightly smaller than an ortolan,' Roy continued. To most of us, in a book or in the field, the two are practically identical.

I suspect it is quite rare for birders who pretty much know what they are about to see a bird well *and for a longish time* and to get it wrong. Real mistakes like that must be infrequent. There have been some of course – female tufted ducks/ lesser scaups, photographed little stints published as semipalmated sandpipers before many people really knew much about telling them apart – but really not so many. The problem comes with the bird that dives out of sight into a bramble bush and never comes out, or a seabird that flies by never to be seen again. We would all like a second look – just to be sure. To be *reassured* – a second look does just what the word says, it re-assures. It gives a chance to check, to rethink, to be quite sure that it did look just as we thought (or hoped?) the first time. Even if the identification remains the same, a good second look very often modifies the perception of detail. Sometimes, it corrects a mistake that a single view might have perpetuated. Sometimes it corrects an error and proves a genuine rarity.

There are other scenarios that are perfectly understandable and not really in the realm of mistakes – some high-profile rarities have been thought to be something else, at first, and the wrong diagnosis has persisted for a day or two. But really that is a matter of improving the perception of the bird, reviewing and researching the possibilities – the identification evolves. Nobody can complain about that. Yet people do, or did: rare birds can bring out the worst in people.

Sometimes, though, like many of us if we would but admit it, I get the jitters. Like a golfer's yips. Maybe I've seen a young ring-billed gull, then, a week later, return to the scene and see a young herring gull looking surprisingly similar. Then I think, could I have... no, surely not – but what if I did? Had I got it wrong last week? Then I remind myself that I was perfectly aware of the herring gull problem a week ago, and had probably not improved all that much since then – and if I can see very well that it is an obvious herring gull now, I would no doubt have done so then. So no need to worry. But, just now and then, doubts creep in – a useful reminder, giving pause for thought.

Not many of us are Dennises or Vinicombes after all.

***

While finalising this book, I did ask Roy Dennis the question. His reply:

> I was more concerned never to claim or publish a record that was not 100% correct because I had worked very hard to attain my field skills and my peers' complete trust – my reputation. I found an aquatic warbler when I was 15 or 16 and earned my spurs. From people in Hampshire I learnt how to identify species by jizz, flight, flock sizes etc. with poor binoculars, hopeless scopes, no camera – just a field notebook.
>
> From the age of 13 or 14 every new bird meant borrowing my scoutmaster's *Handbook of British Birds* and devouring every single word. After school I was assistant warden at Lundy and then Fair Isle, returning there as warden from 1963 to 1970. I was in the field all and every day from 18, and had watched birds most days since 14. So by the time I saw the Cretzschmar's bunting I had watched dozens of ortolans at Fair Isle. They were the norm. I had worked hard to be that capable but I did not think I was infallible. I just knew I was correct – to me it was a different species even with a first look.

FLIGHT LINES OVER THE HOUSE

A sudden explosion of rook and jackdaw calls used to announce the daily movement of birds going to roost over the house. Seeing them in the morning, going back out to the fields, was less usual.

> At about 7.30 a.m., still very shady, I thought it had suddenly come on to rain heavily, but the sound was made by the rooks and jackdaws leaving their roost – thousands of them in a concentrated departure. This afternoon I decided to try to estimate numbers returning, but the first flocks were coming from various directions and I may have missed some, and the arrival was slow and erratic. Nevertheless, eventually flocks of up to 500 or more appeared and the total was well in excess of 5,000, probably 6,000 – several flocks predominantly jackdaws. As usual the main direction was from the north. It was not nearly so good as the best night a month ago, with no mass 'eruption', but still quite a sight. Some flocks fly over directly, jackdaws seemingly going very fast but not making much more progress than the leisurely rooks; some individuals flying over directly but several minutes behind the previous flock; other individuals soaring and circling; and often flocks, or parts of flocks, circling around and being passed by others, often the front or middle of a flock deciding to circle several times and then latching on again to the back of the flock that flew directly by, either 'through' or underneath.

This spectacle continued for a couple of years but has since dwindled; rooks and jackdaws over the house are still a big feature, but nothing like so dramatic as they were in one massive movement. I don't think numbers have necessarily changed much, but the big flocks must be feeding or roosting elsewhere and taking different routes in the evenings. But I do enjoy watching the rooks, and there's always a chance of a raven or two.

Similarly, just for a matter of weeks rather than months or years, we were on the regular evening route of a flock of greylag geese and a separate flock of Canada geese. They would come over, very low, sometimes with a real clamour, reminding me of wild geese farther north, but usually a subdued, erratic chorus of calls. But as soon as I got used to their appearance, they stopped. Wherever they had been feeding must somehow have become unsuitable, and they evidently moved on. More recently, some have returned, if a bit unpredictably.

Each evening, though – sometimes beginning in the early afternoon – from late summer to early spring there is a continuous flow of gulls heading towards the nearest roost. I've yet to see anything rare in the long, leisurely chevrons and V-shapes, mostly herring and lesser black-backed gulls, although Mediterraneans appear now and then at other times, particularly late spring.

The gulls make a fine sight – 'poor man's geese' perhaps, but their long, sinuous lines across the colouring evening sky look good. Sometimes they all momentarily flash white as they tilt over and sweep across a dark sky of heavy cloud. I can't imagine Peter Scott paintings of gulls against the sunset, like his evocative ducks and geese (which were pioneering at the time, despite becoming a bit cliched later), but why not? In winter they come over in complete silence, as a rule, unlike the noisy groups that wander about high up during the summer months.

Why so silent? I've recently read complex explanations why geese call so much in flight, apparently organising their positions within the 'V' to take advantage of the aerodynamic benefits, and give tired or weaker birds a chance of a break. Why, though, do geese call so much when they just skip from one field to the next? Why, when a hundred or two greylags and Canadas fly over my house, calling madly, does the odd one a couple of hundred yards behind try so desperately to catch up, in silence?

Silence somehow adds something, rather than taking it away from the elegant, relaxed gull flocks: a strangely moving daily occurrence, which I don't tire of watching.

THERE WILL ALWAYS BE QUESTIONS

In these days of DNA analysis determining relationships between birds, defining orders, families, genera, species and subspecies, it is interesting to ponder how birds themselves 'know what they are'. DNA studies reveal heritable traits: so the 'family tree' idea can be worked out by tracing lineages from common ancestors, splitting off into different species. This is more or less easy to understand: quite what the DNA people are looking for and finding is more difficult, considering that their scientific papers are written in some kind of extraterrestrial language that few earth-people understand.

A willow warbler might look exceedingly like a chiffchaff to us, but clearly they have no problem themselves, and there are many other species pairs that are far more alike. Bring yourself down to their scale, and the minute differences may well then appear to be blatantly obvious – as well as being reinforced by a suite of calls, songs and behaviours that distinguish the species one from another.

Yet a small living being that one day has to get up and fly from Europe to Africa, where it may change its feathers and suddenly look altogether different, is amazing. It returns to Europe, where it must then form a pair with a bird of its own species, which may well look quite unlike its own appearance as a juvenile, and in any case may still look different according to its sex.

Often cited as particularly remarkable is the cuckoo. In this case, the adult cuckoo has laid an egg in the nest of another species. The egg hatches and the cuckoo chick, seemingly irritated by its fellow eggs and chicks, will one by one manoeuvre them into a hollow in its back and heave them overboard, so that it then has exclusive access to the smaller foster-parents, working hard to feed it. The juvenile cuckoo is brown and barred, whereas its parents are smooth pale grey and, in any case, it doesn't see its real parents at all, but might think it is a dunnock or a meadow pipit, a pied wagtail or a reed warbler.

Off it then goes, all the way to Africa, having been reared by parents that look nothing like it and which may, or may not, migrate. It finds its own way, as adult cuckoos have long since already gone south. It will stay for a year in Africa, perhaps, but then it comes back north in spring (tricky to tell if you are in the tropics), now nearly two years old, and must find... what? A cuckoo. Instinct clearly takes over and it 'knows' what bird to mate with (although there are no long-lasting pairs). A female may lay one or two or even 25 eggs in different nests – sometimes many more.

> 25 July: I came across a pair of meadow pipits feeding a young cuckoo. The young cuckoo was seen very well and looked very brown with a rusty tail: so when another flew over nearby, showing a very grey tail, it seemed a bit odd. They were in an area of rough grass with straggling, bushy hawthorn hedges. I couldn't find the first when I looked back, but then it flew in again, too, again showing its rufous tail. They made the typical juvenile cuckoo far-carrying, tremulous *shreeee* calls that seem to be irresistible to other birds in the vicinity, which come in to force food down the wide-open red mouths. But I could see only one cuckoo being fed by the pipits, to be truthful. It was perhaps a bit naughty but I went closer and flushed the juvenile cuckoo, and out came the second: yes, really, there *were* two chicks. They sat within 10 feet and were then fed by just a single pair of pipits, no question. Surprisingly, a pipit would sometimes feed one cuckoo, then fly straight to the next and appear to feed it, too, presumably having kept some food back from the first. More often, one bird was fed on each visit: both pipits would feed one of the cuckoos frequently, but seemed to visit the other one less often. The cuckoos were virtually full-grown, beyond the initial rounded, ragged, round-winged stage, but still dependent on the pipits for food.

So could two cuckoos be raised in one nest? Surely not, and there was no evidence suggesting they had. Do two colour types – one a greyish bird, the other much more rufous – indicate two female parents? If one cuckoo happened to find the other and just decided to hang around, what happened to its own foster parents? Answers on a postcard... (well, okay, Facebook if you must, although I probably won't see them).

While I was at school I ordered my monthly *British Birds* ('*BB*') via W. H. Smith and my parents collected it whenever they went to town shopping. It remains the important 'journal of record', and most things interesting or important about birds, at least UK-wide, will appear in there, sometime or another. It gave a means for ordinary amateur birdwatchers to contribute, through its 'Letters' and especially the 'Short Notes' section, in which interesting observations are written up in brief reports.

Although I hardly knew him, I treasure brief correspondence with the late Bernard King, 'king of the short notes', who contributed more than 100 such notes over many years, a tribute to his sharp observation but also his knowledge of what had or had not been recorded before. There was, and remains, every possibility for an ordinary amateur to see and record and contribute something quite important, quite significant, to add to our knowledge of birds.

Without labouring the point further, it provided me with an interesting question: I remembered something written in a letter (by field-guide author Richard Fitter) decades before, looked it up and found it, and referred to it in an item written for *BB*. I asked myself and the readers whether I might have remembered quite so well had I seen that on someone's blog or somewhere on YouTube. There is clearly information overload, now, so we just can't remember half of what we see and read, but also, the way we find things out does not seem to be making it so memorable as used to be the case. Perhaps it is just not so easily absorbed. Unless, of course, you happen to be five years old, in which case you will assimilate everything.

Territories, colonies and things in between

As you dip your toe into the world of birds and birdwatching you will soon start to read about birds' territories and territorial behaviour. Many species of birds have territories, but not all.

First, a 'home range' covers the area in which a bird happens to wander – anywhere in its appropriate habitat. Within that may be a more distinctly defined territory, which a male or a pair defend against others of the same species. It is typical of familiar birds such as robin, blackbird, wren, carrion crow... but very many species do much the same. A territory may sometimes be defended by a flock. It may be held for all or only part of a year: we usually think of 'breeding territories' in spring and summer but some (such as the grey plover) have feeding territories that they defend through the winter (which is why you might see flocks of dunlins and knots and golden plovers, but grey plovers are more thinly spread out over a mudflat, even if they group together at high tide). Most territories are large or very large and give the pair of birds all they need: a breeding pair will be able to find food for their growing family within the territory. There are, though, small or even tiny territories, such as the little area – just within or beyond pecking range – defended by a nesting seabird on a cliff. A territory within a colony, if you like.

Think of songbird territories that are defended by, for example, a pair of blackbirds that chase off other adult blackbirds (females as aggressively as the males) and later may even chase off juveniles reared by other pairs. They want exclusive use of the territory, because they feed on such things as earthworms that are evenly spread but not concentrated in areas of super-abundance. They need to search and feed and come back again later, rather than try to get worm after worm after worm from a small patch of lawn.

Males sing, to warn others off, to reduce the risk of having to fight. Although it is often said that red-breasted robins will 'fight to the death', there really is not much point in their doing so – being injured or dead has no survival value at all – and the red breast and song are used to *prevent* fighting rather than to encourage it, to threaten and warn others off. Keep out, this is mine.

Other birds do not have territories in this strict sense. Goldfinches and linnets, for example, feed largely on seeds, although they need insects to feed growing chicks (twites, incidentally, feed their chicks on seeds and are the most seed-dependent finches of all). Their food might be abundant in one small area – even a weedy bank or field – and there will be enough to go round for several pairs. It suits them better to breed close to the feeding site, so they don't worry about fighting, they just settle in a loose group. Finches such as chaffinches also need insects for their chicks and prefer to take, for example, green caterpillars from foliage in spring and summer, so they still need an exclusive territory for breeding, but then come together in flocks at other times, when food such as seeds is more concentrated.

Isn't it amazing, by the way, how such birds can survive on seeds: all year round, pretty much, on a crop of seeds produced in the summer and autumn. Life somehow seems so much on the edge, so very easy to knock off the rails with just a small change here and there.

Flocks have their benefits. A nomadic, wandering flock (such as tits in winter) means there are more birds searching, so a better chance of finding food. In any flock there are more eyes to see approaching danger and, selfishly, if there are 50 instead of one, just a one-in-fifty chance that the predator will chase you, instead of a dead certainty. On the other hand, if there are 50 individual birds and one gets caught, the end result for the population as a whole is the same.

Carrion crows are omnivorous: they eat anything that comes along, be it grain, seeds, dead animals, shellfish, worms... they are best defending large territories, within which they can find enough. But they are flexible enough to come together where food is more abundant or localised: I've seen 100 on a beach, hundreds on fields being spread with manure. Rooks, on the other hand, concentrate on worms and grubs, and grain when they can get it, and feed in flocks. They don't need a territory except just around the nest, to try to defend their bundle of sticks (and bundles of joy once the eggs hatch) from thieving neighbours. So they have treetop colonies of large nests close together, with a lot of noise and display and complicated social interaction going on. Pairs are lifelong and give every impression that the partners like and care for each other: after a fight, one will go back to its mate and indulge in a bit of canoodling, tapping bills and rubbing necks. Love? Why not.

At a superficial level, it is not always obvious why one species will be territorial and another not; why one will nest in a colony, another solitarily. It will, though, be about competition for resources, the costs and benefits of being close to others and working together, or being on your own.

It is easy to become anthropomorphic and allow birds emotions such as our own: but maybe it is too easy to reject these, when there might just be some vague stirrings in the birds' breasts? Humans need to look after their children for a number of years before they become independent, so a strong pair bond for several years at least is 'the norm' and generally a good thing. Many birds have monogamous pair bonds lasting for a season, so they will stay together, work together and help protect each other over the course of making a nest, laying eggs and incubating them, feeding chicks in the nest and then looking after them for a bit afterwards until they become independent. Maybe on to a second or third brood. So a fairly long-lasting pair bond is necessary. Some, such as mute swans, will pair up for life and give every impression of being fond of each other while driving off intruders: but this of course is just 'pair bonding' and defending an all-important territory. Is there the slightest hint of 'feeling' for each other? Probably not. But watch a pair of woodpigeons (famously billing and cooing) and it is easy to think there might be. The smallest bird will fight like a lion to protect its chicks: a biological requirement, but what 'feeling' or emotion inside encourages it to do so?

Since writing the above I have read a book about cranes (*The Call of the Cranes*, 2022) by a German observer, Bernhard Wessling. He faces some of these questions directly, understanding that his ideas might not have been well received in the past, but asking how else to explain what he has seen? I say 'observer', because he looks, as much as researches: he looks and gets to know his birds inside out. Purposeful stuff, indeed. And a lot of what they do suggests to him that they have far more self-awareness than people give them credit for: cranes, he says, 'are individuals that actively engage with their mates, neighbours, competitors and environment. They observe, they decide, and they develop and change their habits.' A 'simple' example serves as an illustration. Area *A* has many breeding cranes; area *B*, 100 km west, the area that Bernhard studies, has a few. Each spring, flocks move north (from southern Europe or North Africa) to area *A*; do pairs from *B* move separately in small groups, or join the larger flocks on a journey to *A* and then turn west? Observations have shown large flocks moving to *B*; from the middle a pair will suddenly detach and skydive down, legs waving, to their territory. The rest of the flock continues at high altitude, but now turns east towards *A*. How did the pair know which flock in the wintering area would take that costly, apparently wasteful, extended route? How did the majority in the flock 'know' they had to make this diversion, drop off a pair or two, then turn for their summer home? With intensive study, including audio analysis, a host of such entertaining, but difficult, questions are posed in Bernhard's book, and a few answers become clear.

One annual survey I used to contribute to, being purposeful myself for a few years, was the BTO's national heronry census. This count of the country's breeding grey herons began as a one-off in 1928 and has continued since as an annual measure of the heron population, with its ups and downs largely related to the severity of the winters.

I had watched a number of heronries from time to time, some mixed with cormorants in later years, but my local one was a large colony in a belt of mixed coniferous and deciduous trees. It is an interesting colony to monitor, as it is close to some big trout farms where RSPB researchers long ago tried to work out the best way to make fish farms and fish-eating birds survive side by side without persistent persecution of herons and cormorants. Tricky.

Nests were in little columns one above the other in several larger trees, scattered through the transparent treetops either side, and in several outliers where four or five together would make extra tight little clumps. Now and then there would be a little egret or two around.

There were herons around almost all year, but activity in February and March began to intensify: herons make a great variety of sounds at the colony, with various snorts and squeals and harsh croaking notes, as well as bill clattering. They also show how well they can fly, as the big, arch-winged birds may arrive from a considerable height and come gliding down, frequently with necks fully outstretched in an un-heron-like manner. Later, after the young had fledged, there might be 10 or 20 or more flying up from a ditch beside a field, one of the old riverside water meadows not far from the colony. The nests were anything from massive armchairs, built up over many years, to new, small, fragile platforms. It was hard to determine just how many of these were really occupied, although herons would usually stand on them, but the nests increased towards the 100 mark in some years. It is always best to watch from a good distance to avoid disturbance, but a heronry is always worth a few hours each spring.

You will no doubt find other things to question, as do we all.

Bird song

... or why yellowhammers refuse cheese, and other mysteries

Like light (don't get me started), sound is both easy to understand and impossible to comprehend at the same time. It is a bit of a weird digression, but relevant to birds in a way (and shows how watching birds sparks off so many tangential thoughts). We can hear them, try to locate them, realise there are flocks of birds calling half a mile away or a chick wanting to be fed cheeping from a bush close by; we can tell one from another by the quality of the note.

Listen to a bird singing, and there is little clear differentiation between individual notes in many cases: most bird calls lack obvious hard consonants. Even *cu-koo* is really more like *uh-oo*, but we need the hard endings to make sense of things. We think a blackcap calls *tak*, but it is really more an abrupt *a*, which doesn't help much. But who knows, reduced down to their scale and speed of action, what a bird might make of it?

The remarkable thing, surely, is the consistency of sound within each species, just like the general consistency in colour. With persistence and a bit of application, it becomes easy to tell a linnet from a greenfinch or a redpoll; to tell a chiffchaff from a willow warbler; a dunlin from a sanderling. More or less every individual of any species sounds *just like all the others*, but different from other species. That is, quite incidentally, handy for us.

Birds sing for two main purposes, it seems: to attract a mate and to defend a territory (or, in bird-book speak, to *proclaim* it). Sometimes, the territory might be a winter feeding area, not necessarily a breeding site.

It is just as well for us that it doesn't, but you might wonder why a blackbird sings so musically (to our ears, anthropomorphically speaking) and doesn't just sit on a high perch and shout *Oi!* The nearest to that might be a bittern, which stands out of sight in a reedbed and calls *woomp* as loud as it can, or perhaps a carrion crow with its raucous cawing.

Recently I spent some time in a car park by a couple of warehouses and a 'retail park', a handful of straight-walled, square, flat-topped shops. Don't ask. It was late July, well past nesting time, but the local herring gulls and a few lesser black-backs were whooping it up. Niko Tinbergen and his students long ago studied herring gulls, and I think he said something along the lines of the herring gull having the most dramatic and extraordinary bird call in Britain (or wherever). The 'long call' is essentially a territorial call, defending a small patch from other pairs close by on a sea cliff or, perhaps, a flat roof. But here were these gulls, five of them standing shoulder-to-shoulder, swinging their beaks up to the sky and giving the most spectacular long, loud choruses of bugling and explosive trumpeting sounds that reverberated around the sounding boards of the buildings – truly the most amazing bird sound around. Why?

And if this is a territorial call, why then do passing herring and lesser black-backed gulls high overhead, far inland, regularly give the same performance, which can be heard for miles down below?

The volume of such calls is remarkable. You might not think the same about a warbler's song, but consider. Sometimes you have a job to hear what someone is saying to you across the lawn, let alone from a couple of houses down the road. And yet you can hear a calling goldcrest from deep within a wood as you pass by in a car, or the song of a chiffchaff from a quarter of a mile away. The penetration, and carrying power, of minute calls from tiny throats is unbelievable! What about the ringing *tswee-wee-wee* of a common sandpiper calling from the other side of a reservoir or Scottish loch? Or the *tew tew tew* of a greenshank that echoes across a big estuary? (It echoes unrealistically around every suitable-or-not coastal or wetland location in dozens of television programmes, too. Why do they *do* that?)

Goldfinches sing around my garden, often from the television aerial on the house roof. They are semi-colonial, not really territorial, but sing all the time. But why sing after they have reared their young, with a slightly rambling, unfinished but prolonged song that almost gets on your nerves hour after hour, day after day in summer? And then, one day, they are gone.

There are many such 'why do they do that?' examples that refer to birds, not people, that arise when you watch birds for more than a few days.

On one hot day recently, I watched a robin hop out from some shrubs onto the edge of the lawn, and promptly squat down, wings open, tailed pressed down, almost fully into the 'sunbathing' pose – and it started to sing! I recall an early morning in midwinter with a severe frost and white rime over every twig and branch, the kind of day when small birds supposedly have to feed and feed and feed just to maintain their body heat and somehow to survive the long, cold night. Nearby was a tall conifer, white with frost; from the middle came the song of a goldcrest. Why did it *do* that?

Bird song usually helps identify a bird with little difficulty once you know it. Some species are not so easy: garden warbler and blackcap, for example, especially if they imitate each other! Ex BTO Director Jim Flegg noticed that these two in woods in Kent would also try to copy local nightingales, as if the 'better' bird was pushing them to higher levels of expertise, and in these circumstances identifying blackcap and garden warbler by song – normally not quite such a stiff test – became well-nigh impossible.

The best way to learn a song is to watch the bird that is singing, as it sings, so you can match sight and sound conclusively and fix them in your mind. It is harder listening to a recording of something that isn't there at the time, still harder reading descriptions in a book. Written descriptions of songs and calls are there to help people remember, rather than to imagine something as yet unheard.

The yellowhammer has one of those distinctive rhythmic songs that lends itself to a written phrase, long interpreted as 'a little bit of bread and no cheese' (although that doesn't quite get it, in reality). You can't do that with a rambling, prolonged song from, say, a robin or a blackbird (although you might spell out phrases spilled out by a song thrush, which has a sizeable repertoire of notes that are each repeated several times over). One person's 'rich, melodious, musical' might be another's 'throaty, bubbling, repetitive' but a general idea can usually be conveyed in words. But there are now websites with extraordinary collections of bird calls and songs, for practically any species you come up with (try the exceptional Xeno-Canto), and comparing what you hear with a recording is easy. The problem is, if you don't know what it is you are hearing, you don't know where to turn. But increasingly there are apps for smart phones and the like that will identify a bird call for you. Too easy.

Calls, rather than song, are also extremely helpful in many cases. A simple example is that a willow warbler or chiffchaff calls a *Phylloscopus hooeet* while a garden warbler or blackcap calls a *Sylvia tak*. Listen closely and the willow warbler calls *hoo-eet*, distinctly double, while the chiffchaff calls *hweet*, more nearly monosyllabic; and the blackcap has a hard, tapping *tak* while a garden warbler's call might be just a touch softer. Start easy and delve deeper into the more tricky ones. But the calls of, say, a marsh and willow tit can be the best way to tell the two apart, so getting into bird vocalisations is important. Do your homework.

Sounds also help in other ways. You will soon learn to recognise the alarm notes of a worried starling, and can look up to see a passing sparrowhawk or hobby that you might otherwise have missed. The noise of a group of small birds all calling in alarm around a big tree, perhaps a holly or evergreen oak, can get you close to a roosting tawny owl. The remarkably distinctive call of a Mediterranean gull can draw attention to one flying overhead in spring, almost anywhere.

Anyway, helpful or not, bird calls and songs are simply enjoyable and, as with our other senses, sound has the ability to transport us to another time and place entirely. In 1994 I see in my notes in Cambridgeshire '*Skylarks, SN – excellent song, still enough to take me to a windy Scottish coast or Welsh hilltop!' Or it may be back to childhood, or far away to some exotic holiday destination.

Many years ago, I used to watch the gulls in Swansea Bay. The local gull and wader watcher, Bob Howells, had been watching and counting roosting birds for years, but at the time I was less respectful of his expertise (and sheer hard work) than I might have been. At that time, Mediterranean gulls were beginning to turn up in southwest Britain and the first ones had appeared in Swansea. When we saw our first it was Peter Garvey who told me what that strange-looking immature gull was. Much later, when I mentioned that I'd seen one, Bob said yes, he'd heard them calling. Ha! How do you tell one gull from another on a busy beach?

Now I know. Sorry, Bob. The strange *yaa-wooo* call stands out a mile.

Female choice and the pressure to evolve

If you want to tell a chiffchaff from a willow warbler, or a pectoral sandpiper from a dunlin, or a booted warbler from a Sykes's warbler, you don't need to know much about any other aspect of bird biology or behaviour, but even so, it is always interesting to delve into things that help you to understand and appreciate what you see: which is what this book is all about. But when you come across nice neat theories, it is also sometimes interesting to stand back and think about them for a bit: do they always stand up? Sometimes things seem so difficult to understand that it is probably best just to accept them: don't worry about it, and carry on.

I've long wondered about the idea of female choice as a selection pressure that helps drive evolution. It is clearly a fact and stands up well to careful and objective observation, but there are also some things about it that are quite hard to understand. By questioning such things, I am always wide open to criticism and show my ignorance... but never mind.

If males display at a lek, that is in groups showing off in ritualised displays while being watched by secretive females on the fringes, it is easy to see that the female will, or can, choose the best mate – the fittest, revealed perhaps by the finest colours and plumes, the most impressive behaviour. Even Darwin, however, thought that not all such things necessarily revealed the 'fitness' of the male, and that birds displayed beauty for the sake of beauty, or for beauty to be appreciated for its own sake. American ornithologist Alexander Skutch also, I believe, thought that beauty was there to be appreciated for its own sake. This idea, ignored at the time, has recently gained more ground. Females don't necessarily choose the best-looking, best-sounding, or top-performing male because he will be the fittest, but for aesthetic reasons.

Still, given a range of males, the female can choose the 'best' for whatever reason. Recently, however, I heard a radio programme about nightingales, and the female-selection theory was expounded as ever... males continually develop and improve their songs, so the female can judge by the perfection and complexity of the performance which male is the 'best' as a potential mate. I contacted a renowned expert and queried this and was told 'Your problem is, you don't understand.'

Imagine a small suburban area with gardens and parks, full of blackbirds. Or, say, picture a row of ten male blackbirds with the 'best', or fittest, somewhere in the middle. And a row of ten females lined up underneath. So the females have a choice and will pick out the best male, by his song and/or other signals. Okay. But hang on a bit. How do the females sort out which one gets the best male? There does not seem to be any complicated display system by which the females sort themselves out. So, any old female might get the male: but in any case, as a rule, more or less, all ten females and all ten males will pair up and get along fine. Sometimes you might hear something like a wood warbler singing away all summer because there is a shortage of females, but generally, roughly, on average – they all find a mate and seem to be happy enough with what they've got.

And while a nightingale or a marsh warbler, say, can develop its song to give a powerful indication of its fitness as a father, a chiffchaff or a goldcrest, for example, will chunter away year after year with the same song as all the others – yet the females fancy them, just the same.

The 'best' male might produce fractionally stronger chicks. As Darwin pointed out, to understand how this might be important and effective in evolutionary terms, we have to get used to thinking in terms of millions of years: you can see that millions of minute improvements will add up, eventually. But the thing is, surely, that the best male would have produced the best chicks, whatever female he paired with: unless there was also a 'best' female. And how do they, and he, know which one that is? And does no. 1 male pair with no. 1 female, no. 2 with no. 2 and so on down to 10? If the females were sorted into some sort of league table, and there was no female selection at all, the result would surely be exactly the same.

Swallows move around far more freely beyond their nesting areas, and it is well known that many young swallows are fathered by males other than the one that the female is paired with – loads of extra-pair copulations. But females are supposed to choose the best males according to such factors as the length of the tail streamers. Does one female somehow dominate all the others to get the best male? Does the best male slip off while his mate is incubating their eggs to have a quick affair with the other females? (The female must be pleased she chose him, then.) Do top-rank male blackbirds also put themselves about a bit to improve the local stock? Do females from other pairs encourage him? Do they actively seek a quick affair with the top male (female selection in action) or do some just occasionally see him coming and mate with him at random, simply acquiescing (no female selection at all, then)?

Tropicbirds display around mostly smallish colonies: there might be 20 pairs, say, all monogamous, so 20 males paired with 20 females. The females apparently select the male with the longest tail, shown off in his display. But which female? Which of the 20 gets him, and how? And then you read that the pairs are also neatly assorted, so the male and female of each are well matched in terms of tail-spike length. So there is male selection as well as female choice going on? Or what? How do they do it, if the males show off, revealing which is the best, but the females just look on, passively? Either way, tropicbird display is supremely elegant and exciting for us to watch.

Okay, so the 'satellite' white-feathered ruffs will display at the edge of the lek and draw the females (reeves) in, while the 'best' dark-plumaged males will get the most hens; the brightest, strongest, loudest, longest-tailed pheasant will attract the most females. But ordinary everyday territorial songbirds? It is not so easy to see how it works.

One relatively new discovery, however, is that almost all birds use their sense of smell much more than was once thought, and, not only that, some – especially females – produce special scents for social/mate selection purposes. There must be something there to help my foggy wonderings

Unless you can see where I'm going wrong, which is *very likely*. Don't worry about it, either way.

Evolution is astonishing: so astonishing that even if I have not the faintest clue how it works (as above) I can sit back and admire it as a wonderful thing.

I mentioned the idea that birds may select their mates not for the 'fittest' but for the most 'beautiful' candidates. This idea has been thoroughly researched and developed by Professor Richard Prum, whose writing is illuminating, informative and provocative as well as sometimes annoying because it leaves me feeling frustrated and ignorant. He particularly describes the sensational plumage and displays of the male great argus, like a big pheasant, which he rates far above the peacock in terms of adaptation for pure aesthetic appeal. It is phenomenal. Yet how can it be? I can't easily see how this remarkable bird can have evolved simply through female choice. It has elongated secondary feathers that have a complex pattern of stripes and spots and a series of golden rings around 'shiny' discs: these have an interior patterning of a black outer ring that blends into a beautifully graduated dark brown and white to create a fantastically realistic impression of a three-dimensional sphere, the white being near the top but just offset, like a glint of reflected sunlight. Also, he suggests, females must like a kind of perfect symmetry because the farther along the feather, the bigger the sphere becomes, so from the point of view of a female looking at the displaying male head-on, with these feathers raised, the spheres all look the same size.

It is indeed fantastic: you can see it all easily enough on the bird. But how can some random mutation ever have produced such a perfect 3D globe: a bit more shadow here, a bit more highlight there, a bit more of a glistening effect at the top and depth of colour towards the bottom, all nicely, smoothly graded like a shiny metal ball. How did it begin? Let alone how it gradually improved as females demanded more perfect shiny globes. Remember to think in millions of years, millions of generations: but it is easier to understand the gradual development of an eye complete with optic nerves and the rest than it is to see how this feather came to be as it is, driven by female choice.

I am not anywhere near being a creationist, but this kind of thing sometimes takes people more closely towards some kind of 'intelligent design' than they are comfortable with. There is a natural tendency to push questions aside and get back to more obvious examples. Once I wrote a short spoof story about the evolution of ear wax. Hmmm... here's a shorter than short version. Ear wax is a mix of perspiration and skin fragments. Now isn't it remarkable that such a thing should evolve and lodge itself inside the deep, dark, damp recesses of the earhole? Where it oozes out and cleans dirt from the ear, or sinks back and blocks it to produce temporary deafness (evolution hasn't finished, yet: it might get better). How come it didn't first appear behind the knee, or under a big toe nail, or under the left armpit? How did it just happen to start up in the one place it might (or might not) be useful? It is easier, maybe, to think that early 'man' was covered with this stuff and it wore off everywhere else until only the hidden bits in the ear remained.

You can see that none of this is meant to be any kind of serious biological debate, and anyone can pick holes in my slightly flippant and superficial arguments, but it does show that an enquiring mind might lead to more interesting things even than the closed wingtip length of a melodious warbler. How on earth did an earhole evolve, later to be equipped with an eardrum (was it any use without one?), all the little bones and hairs and whatnot that make it so effective? That control your balance, too. How did mammals all happen to have blood complete with red and white cells, platelets and immune systems and goodness-knows what else, which surely cannot have evolved multiple times but must have been there from the beginning, in the prototype mammals of 200-odd million years ago? A mouse and a whale both have much the same as we do.

It is often easy to understand a difficult question once someone gives you the right answer – well, obviously. Recently I was writing about hummingbirds. Why would a flower evolve a very, very long, very narrow, tube-like shape so that it could be pollinated only by a hummingbird with a very long, straight bill? Would it not be better to be short and wide-open, so that any hummingbird, with any kind of bill, could come in and take the nectar and carry away the pollen? A moment's thought – hmmm, tricky – then I was given the instant answer. Of course, the flower needs that hummingbird to go away, not to any old flower, but to *another of its own species*. So, the bill-shape/flower-shape combination, an example of co-evolution, forces the hummingbird to go away and find another flower *of the same species*, so bingo, pollination takes place. If it could feed on any old flower, the chances of pollinating the same species would be much reduced.

Obvious. I think.

UP ON THE MOORS

Going through my old notebooks brings back so very many memories and experiences, some of which simply reinforce what we have lost. There is much that we have gained, too, but in my early wanderings around local lanes, turtle doves, tree sparrows and willow tits were just 'noted', in the usual 'small numbers'... they scarcely merited a second thought. They have long gone.

We also used to have regular visits, especially around Easter, to the beautifully picturesque moors in north Staffordshire. For a start, being up on top of the world was a great experience in itself, with views over Cheshire and back south across the lowlands of mid-Staffordshire, far away into the hazy pale blue. There were little hidden valleys and boggy triangles, and exciting areas of green pastures outlined with tall ash and scattered birch trees, splashed with vivid yellow dandelions (the best!), often close to an old, inconspicuous, grey stone farmstead. There were well-known high rocky outcrops and crags such as The Roaches, where few people then bothered to go. Breeding ring ouzels were left undisturbed; golden plovers sang from surrounding wide, round ridges. Such places are dotted with climbers, hikers, runners, mountain bikers and the like, now, enjoying these great areas as people should, but much to the detriment of the ouzels, curlews and golden plovers, short-eared owls and merlins...

One bird that we watched as often as we could on these trips was the black grouse, now altogether lost from these moors, a really sad situation. There were red grouse widely but thinly distributed through the heather on the heather; lines of shooting butts always jarred. It always seems a great pity to me that they are just thought of as 'game birds' and hardly get the credit they deserve as wild (not reared-and-released), beautiful, evocative birds of the hills, with their brilliant calls. Bird sound does not have to be conventionally beautiful to be special: think of the cackling of a fulmar on a Scottish cliff above the surging waves. The staccato phrase of a red grouse echoing across the rolling slopes helped make these upland moors so very special. W. H. Hudson, mentioned several times in this book, described the male as like a bird carved from some specially hard, dark, red granite – and so indeed it is, with its vivid red wattle over the eye to set off the glossy plumage.

The black grouse were more restricted to a handful of well-separated areas, mostly with just a few pairs. Once fragmented like this, the habitat and the grouse were clearly heading the wrong way, no longer sufficiently robust to resist change and environmental pressures. In one valley there might be golden plovers singing somewhere, ring ouzels on the grass, wheatears on the stone walls, a black grouse or two, and a couple of red-necked wallabies... these 'escapes' lived

Black grouse have more recently become more accessible to people at carefully organised watch points: look at RSPB events on the website. Seeing them is a very special, out-of-the-ordinary experience.

in the area for many years but have also now gone. There was one larger lek where the black grouse gathered to display, which could be watched from a distance or from an old, empty stone barn (recalling Private Frazer – 'did you hear the story about the old empty barn? There was nothing in it.'). I roughly sketched a few poses.

We didn't see full-on leks (probably visiting at the wrong time of day) and didn't disturb the birds, but days out, up on the moors, were exciting, with so many singing larks, the ecstatic bubbling of curlews, those great challenging bursts of calls as red grouse exploded from the heather, and always the chance of a short-eared owl hunting by day, a pale, sandy apparition over some distant dark heather. And simply a very different landscape from that closer to home, from the high tops to the clear cold waters of the Dove, with dippers, beautiful brown trout and graceful graylings.

But it was the black grouse that topped off the list. Males on a field of short green grass can be seen from a mile away, and heard almost as far, sneezing and crooning like pigeons. The Staffordshire ones would often glide down a shallow valley, against the background of an old stone-built farm with its sheltering trees, their bold white wing-bars shining against the blue-black. They are fantastically handsome, seemingly too exotic for an English bird, black, glossy steely-blue and brilliant white, just neatly topped off by a touch of red over the eye.

Most pictures, whatever the subject – photographs or paintings – are enhanced by a little touch of red. Look at Constable's.

GULLS

Gulls are brilliant, honest...

Gulls are not everyone's cup of cappuccino, I admit. Some people hate them. Some birdwatchers dislike them. Many ignore them, if they can. For me, they are endlessly beautiful to watch and fascinating to study, although as time went on it became ever more clear that you needed a degree in gull-ology to understand them and identify them. By the way, I always call them gulls: 'larids' is an affectation I can do without.

When I walked time after time around my local reservoir, Chasewater, I began to notice that the gulls were coming in as I made off home. It occurred to me that I might wait a bit and see what went on. I don't think anyone had told me to watch the gull roost, although plenty of other people were doing it (though, to be fair, mostly at other reservoirs such as Belvide). I once counted 450 lesser black-backed gulls with four great black-backs in the middle. Looking at my books and reports I noticed that 450 was an unusually large count: and I pointed out the great black-backs to a much more experienced observer, who said he hadn't noticed them. There was potential, here, for finding something new and different. (One day many years later I counted more than 500 great black-backs, still extraordinary, while lessers often exceeded 2,000.)

Now, most gull roosts are well watched and recorded, with more or less accurate counts or estimates through the year or from autumn to spring. Gull-roost watching has become a national sport. It is often cold and uncomfortable: even a hide can funnel in freezing winds, drifting rain and snow – be prepared for your hands to freeze onto the telescope. But it is *great*!

Gulls are mostly grey, white and black and so are particularly affected by the variations in light and weather conditions. They can look dazzlingly white on blue-black water, or dark, grey shapes against a slab of calm, silver water, or shimmering against the sparkle of innumerable ripples catching the low winter sunshine. In strong low light they may look blue and orange, like a painting by Cézanne. In falling snow beneath a lead-grey sky they create a perfect colour chart, a smooth gradation of white and grey tones, when everything becomes clear, uncomplicated by strong highlights and deep shadows.

When people were working out what different subspecies of herring gulls were around, and also showing that yellow-legged gulls were regular in England, some of my notes were criticised by researchers who were more expert in museum studies than in field observations. They pointed out that judgment of back colour was very hard as herring gulls might turn and face in different directions, so some would be brightly lit while others were in shadow. *Duh.* I'd never thought of that. Anyone seriously watching gulls at reservoir roosts would be very well aware of the pitfalls.

There are four or five regular species in most areas: black-headed, common, herring, lesser black-backed, maybe great black-backed gulls. In coastal areas maybe kittiwakes. At an inland roost, these are the basics, the ones you will see most, in hundreds or thousands. Over the years, at gull roosts on reservoirs and gravel pits, I have seen 15 species, with a couple more in flocks at the coast. In addition, several well-defined subspecies and a few unknowns. Try hard and what might you find?

As I'm not a ringer, or a photographer, or much of a museum worker (I've looked at just a handful of museum gull specimens), my observations might be a little bit inconsequential. Most, now, require good close photographs and detailed analysis. Usually, I've watched gulls at long range on a lake, while it was steadily getting dark, or else on a wide beach as the tide came in. Getting all the detail really needed was usually difficult.

Many of my early notes now seem a bit weak: this or that gull, with a vague idea of its appearance, a query as to what it might be, then it is left. The thing to remember is that there were then no special books on gulls; nowhere really for an ordinary birdwatcher like me to go to find out more. Only rarely did I visit a museum to examine skins. It was not just that there was no easy information on the possibilities, but we just didn't even know what those possibilities might be. No Gull Research Organisation website, no websites at all. No internet, no Wikipedia... nothing. It was a while before Peter Grant mentioned to me that he had a photocopy of Dwight's 1926 epic 'Gulls of the world: their plumages, moults, variations, relationships and distribution', a modest little 300-page paper in the American Museum of Natural History's *Bulletin*. Somehow I got hold of a copy.

'That looks different. I wonder what it is?' might be as far as we could get way back when.

*In 1971 the first **Franklin's gull** for Britain (from North America) was found in Hampshire and, making a somewhat adventurous hitch-hiking outward journey, I went to see it. Some years later I also went to see the third for Britain, in Suffolk, and one of my drawings copied from notes made while watching it is shown here. Years later again, I found one of my own: I've been lucky with gulls, but, as Gary Player said of his golf, the more I practise the luckier I get. I have looked at millions of them.*

Nevertheless, Midland gull-watching was illuminating, especially in terms of sorting out yellow-legged gulls and subspecies of herring gull and trying to see the differences between glaucous and Iceland gulls. Some winter herring-type gulls were obviously darker-backed, much whiter on the head, and had yellow legs. Now, these are clearly yellow-legged gulls *Larus michahellis*, separated as a different species (which replaces herring gulls in southern Europe) in 1993 by some authorities, but not until 2007 by the British Ornithologists' Union (BOU). In the early days it was hard to get the 'authorities' to take things very seriously, as head colour, back tone, wingtip pattern and leg colour are all individually variable to a degree. It just seemed that to have all of these looking unusual on a single bird seemed a bit too much of a coincidence.

The herring gulls were either 'ordinary', pale and streaky-headed, or bigger, darker-backed, often with much less black in the wingtip, with long, dull bills. These it turned out are of the northern (Scandinavian) subspecies of herring gull, *Larus argentatus argentatus* (British ones, smaller and paler, are *L. a. argenteus*).

Then there were lesser black-backed gulls, and a very few that looked blacker on top. These we called 'Scandinavian' lesser black-backs because that was the best we could do from an ordinary field guide: that's what they called them. They were, however, likely to be of the northwest European subspecies, which can be darker or even blacker above than 'ours' (the palest, *Larus fuscus graellsii*) and sometimes whiter-headed in winter. But 'Scandinavian' ones, particularly Baltic birds, are different and often now separated as a different subspecies, Baltic gull *L. f. fuscus*. We don't get them often, because they head off southeastwards in autumn, not southwest. I've seen the odd one, on the Danube Delta. The dark-backed birds (of the subspecies called *intermedius*) were very scarce in the West Midlands when I was watching there but proved very much more frequent in recent years when I watched gulls in Hampshire.

There were, though, many troublesome individuals – and, far away in fading light, they remained uncertain. So I would often see something smaller than a typical herring gull, therefore smaller than the rather darker yellow-legged gulls, but clearly darker above. Small, slim, dark-backed... what were they?

In Hampshire, lesser black-backed gulls included the big, dark ones, but year after year I would quite easily pick out a handful of individuals – conceivably the same each year – that looked a bit bigger than the British lesser black-backs and just a tiny bit *paler*, a little more milky-grey above. Other people called them 'hybrids', presumably meaning herring × lesser black-backed, but I always felt this was a bit of a cop-out. Maybe. But who really knew?

Some notes from frustratingly inadequate views:

> Pale lesser black-back type: recent ones have looked smallish and sleek, with one or two at least showing small, more or less pointed white tertial patch. Today's was not a small bird, about typical *graellsii* size with a rather broad and deep tertial crescent giving a more substantial look. Head moderately streaked, nothing unusual in weight or pattern or marks, bill not large nor specially small, just a dullish *graellsii* type. Upperparts a clearly

> distinct slightly paler milkier grey – lesser in low sun often take on a slight gingery-brown tinge, which these birds possibly lack. Wingtips of previous birds looked long, pointed, black with large white spots reducing towards tip. This one had short wingtips, black with large white areas, seeming to be a white patch with a black spot, almost as if a mirror was revealed (not sure how this could be). Not seen to fly or open wings.

A vague possibility might be *Larus heuglini* (or *L. fuscus heuglini,* depending on whether you consider it a species or a subspecies of lesser black-back). It comes (if at all) from Siberia.

> 'Heuglin's/Siberian gull' – 1 adult – a good candidate, this one, like average *graellsii* in size and build, with similar closed wingtip possibly still a little short, but also hunched and sitting low in the water. Slightly but again distinctly and consistently paler, in varying light conditions. Head pale with faint streaking but more distinctly dotted hindneck. Bill quite bright yellow with blackish arrowhead/diamond shape close to the tip (a feature labelled 'typical' in Lars Jonsson sketches of Heuglin's). Good views but just a little distant and never once even raising its wings!
>
> ...
>
> Soft misty conditions at first: lesser black-backs variable but difficult – only one or two obvious dark *intermedius*, but one or two just slightly paler than *graellsii*, looking long and slender, smaller than yellow-legged, with streaky head and breast – odd things that I could make nothing of. Then as the light improved and the air cleared, and more gulls arrived, there seemed to be more quite distinct *intermedius*, some very dark at least from some angles, one or two as usual looking small, low in the water and slender, but streaky-headed. They are puzzling things.
>
> ...
>
> I become increasingly frustrated by birds that seem to me to be unidentifiable here. I usually see large, fractionally pale 'lesser black-backs' (which, despite the seeming impossibility of identifying them in the UK, still seem to me to be candidates for Heuglin's gulls). Today I saw two distinctly pale birds that were *smaller*; one seen very well, quite close, in varying lights, picked up repeatedly over a long period in different places. It looked adult on the water and on the spit, but in flight revealed dark black-brown outer primaries and some streaks on the inner primaries and primary coverts – not seen very clearly. A small mirror appeared to show

at rest, together with c. 4 small white tips. The distinctive feature was the upperpart tone, somewhere close to yellow-legged gull but a different 'quality' of grey, a kind of steely or lead grey – just, but very clearly, paler than *graellsii*. It looked small and slender, long-winged; head small, round, particularly high-domed and with a steep forehead; bill quite small, dull pale yellow with a brighter tip and red spot with irregular thin black marks intermixed. Eye pale yellowish, with thin ring and no real concentration of streaks around it; head pale with weak, sparse brownish streaking, giving a bland, open, almost peevish expression; but hindneck white with discrete thick dark streaks expanding downwards and out over the lower neck sides in a wide shawl of streaking; also extending as a wash with blurred marks around the foreneck. This pattern is unusual in lesser – which mostly have a lot of streaking higher on the heads but a whiter hindneck. Legs strong yellow. It looked distinctly odd and 'different' but will just not fit into any alternative possibilities. Even hybrid origin does not easily explain it.

...

I can never get any farther with the big gulls at a distance, which are sitting on water and determined not to raise a wing... one looked like a small, slender lesser black-backed but distinctly a little paler, with a small, slim, pale yellowish bill with a dull red spot, and the head slightly dull, softly streaked down the back of the neck but with little marking on the side, and just a small dusky area immediately around the eye. Another looked the same shade above, somewhere close to yellow-legged but a fraction darker, always paler than a *graellsii* lesser; a little bigger, with a bigger, rounded head, bill held angled down, dull, pale, with weak red spot; dull head with a little soft streaking behind, more on the lower side neck, not much around the eye. These could be almost anything... I could not be sure of the eye colour of either but perhaps not obviously dark, and I did not see leg colour.

A complete range of upperpart tones: some particularly pale-looking herring gulls, many 'normal' ones, and some darker ones, some candidates for *argentatus* [*the big, dark north European subspecies of herring gull*].

Then a bird that was darker still, but rather *small*, with a small, weak bill, yet heavily streaked head and especially coarsely dark-streaked hindneck and neck sides, and browner streaks over the whole of the breast – I can only imagine it was a small, dark *argentatus*. However, glimpses of flapped wing at long range made the pattern difficult to judge, but maybe just one mirror – too little white – and probably three outer primaries black to the base, seemingly too much black for *argentatus*. There was at least one other similar one but with only the head and neck more weakly streaked, and a white chest.

Then there were 'dull-headed' yellow-leggeds as well as immaculate white-headed ones. And at least one was clearly darker again, very close to *graellsii* but distinctly, fractionally paler ('milkier') – large, with light head streaking. This would be what I have in the past called potential *heuglini* types, but what might actually be no more than a hybrid of some sort. And then typical *graellsii*, darker lessers, and eventually practically black lessers clearly at the darker end of *intermedius*.

...

More a spectacular gull roost than a good one for watching, although at times very good views in excellent light of many birds, especially the common gulls, quite close up. There was a roaring, blustery gale that faded away a little as I watched, while it became colder – in bright sun, the gulls looked great but were constantly disrupted, I think just by the big waves and strong winds.

...

A great gull roost in strong winds and overcast conditions, paradoxically making the light ideal for gulls most of the time – in the gusty wind, they came in closer than usual, strung out right across the lake, with other tight flocks right over the far side. A great deal of movement making following individuals difficult. And a lot of 'maybes' and unidentified birds. Great to watch, though.

...

Some superb yellow-legged gulls – really fine birds; third years included 1–2 with blackish bills. Two or three seemingly adult with a smudgy area of brightish brown streaks around the eye, the maximum 'autumn' extent.

One bird just paler than *graellsii* lesser, just darker than yellow-legged; not very big; bill not large but quite bright; head not really hooded but upper half streaked; a particularly dark eye-crescent and dark around the eye, and an upswept smudge of streaks behind the eye to the nape; eye looked particularly bright yellow. Wingtips showed four white tips and apparently one large mirror.

> One herring gull striking and easily picked out repeatedly – big, long, low, head held very low, crown particularly low and flat – quite like a glaucous shape – markedly pale, with black primaries and something around the tertials (hard to see clearly), dark eye, and blackish bill with faintly paler tip and base. Other herring gulls small and pale, almost insignificant by comparison, hardly looking like the same species.
>
> Three or four distinctly dark, streaky-headed herring gulls, not especially large, or pale-billed, but presumed *argentatus*, which seems to be rare down here.

In Hampshire, years after my most concentrated period of watching gulls in Staffordshire, we could usually find a Caspian gull or two. Caspian gull is another piece in the herring/yellow-legged jigsaw and became much better known and more frequently recorded after the mid/late 1990s. Now the obvious inference is that better-known (and with more people aware of them) would equate to more frequent. Yet there were enough people in Staffordshire scrutinising the gulls often enough and hard enough that I still feel sure we would have noticed if Caspians were there, even if we may not have had any idea what they were. I can imagine the 'Look at that – what do you think that one is?', but there seemed no hint of them then.

Yellow-legged gulls are usually big, large-billed, with the bill vivid yellow and marked by a red spot: sometimes a black mark too. Now and then in the Hampshire roost there would be something quite odd: a really handsome (or even 'pretty') gull, a little smaller and slimmer than might be expected, with a clear white head against the mid-grey back, a lot of black in the wing, and obvious dark eyes and a smallish, straight-edged bill that looked vivid yellow with a broad black band and red spot. Not so easy to place. Oh dear. Gulls can be frustratingly difficult once you get beyond the basics. But, as I have said already, they are not everyone's cup of tea.

In Hampshire glaucous and Iceland proved to be rare. In Swansea, I saw several of each. One adult glaucous flying above me against a deep blue sky, in brilliant sunshine, remains one of the more beautiful gulls I've ever seen. In Staffordshire, I could occasionally see up to four glaucous and a couple of Icelands at a single roost, which is fantastic for any midland or southern English site. The best idea was to sit in the car, listen to the football, and watch the gulls come in, but much depended on the direction of the wind: gulls swim head-to-wind and a side view is always best.

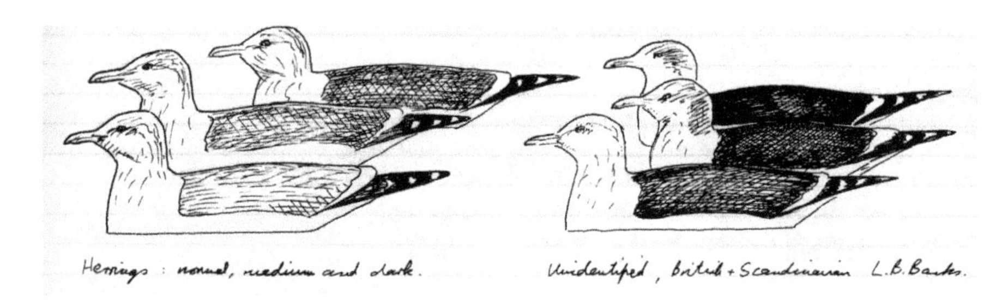

My frustration over difficult gulls has been evident for decades... these are early Staffordshire observations. Too imprecise, but they were miles away in gathering gloom. 'Scandinavian' would presumably be intermedius, *clearly not* fuscus. *(Have you noticed that gloom always gathers? Like clouds that always scud.)*

But some have worked out well enough. This sketch goes back to mid-1970s, when few people really bothered about potential yellow-legged gulls, or just thought of them as individual variants. The right-hand bird was what now would be a perfectly ordinary, typical adult yellow-legged gull, but then was a mystery that most other people either didn't notice or preferred to ignore. There seemed to be no confirmed records of 'other subspecies' of herring gull in Britain at the time, and yellow-legged was not separated as a different species.

If you haven't got a big roost to watch, you can still see a surprising number of gulls even at a town park lake. I've seen two Bonaparte's gulls and two different ring-billed gulls (vagrants from North America) in such situations, during the day, and there are always places such as the car park at Radipole Lake in Weymouth, which has famously had a number of rare gulls gracing its tarmac over the years.

In fact, flocks of gulls, flock of ducks, flocks of waders – all are worth scrutinising if you fancy the idea of finding something unusual. It can happen as often as once every 10 years!

Common gulls

'Ours' are *Larus canus canus*. Far eastern ones are *L. c. heinei*. In western America you get the mew gull or short-billed gull *L. brachyrhynchus*.

> I used to think, when I saw a lot of common gulls years ago (e.g. Swansea Bay) that adults were remarkably constant in tone, like black-headed gulls. These at Ibsley show a bit of variation, with one or two very slightly paler birds (still too much contrast and too much white on scapulars and tertials to be anything like a ring-billed) but more especially a few possibly large birds that look noticeably dark. I suppose there are a few western *heinei* regularly that are not identifiable, or *heinei–canus* intergrades, but these dark ones must be identifiable *heinei* from farther east.
>
> ...
>
> One large dark common gull with a white head. Some others darker than most. One very odd bird watched by several of us – a big bird for a common gull, but clearly smaller than a lesser black-back – it looked long, high in the water, the slim wingtips tilted up high; big head like large, chunky

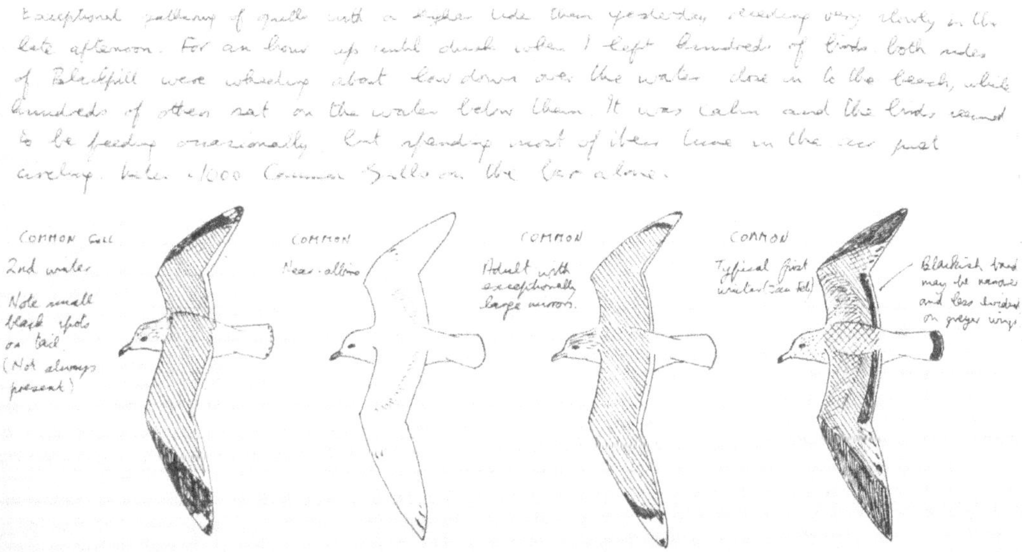

Common gulls: a second-winter (i.e. 18 months old) bird, a leucistic (near-albino) one, a bird with unusually large white 'mirrors' in the wingtip (oh, could that have been a short-billed/mew gull?) and a first-winter, sketched in Wales.

> common or ring-billed type, but bill quite thick and long, and dull yellowish with a big black ring, like a well-marked ring-billed. It is hard to suggest an identification unless it is a big, heavy, dark *heinei*, but the bill looked too big and prominently ringed even for that.

One that got away? It was quickly 'lost' but I wish I had paid more attention. Coincidentally it was on 14 March, a date on which I have seen several rare gulls in the past.

> A very peculiar and eye-catching common gull, that fits nothing... in flight, striking, with wingtip seemingly 'streaky' and lacking extensive black – better views revealed two very large white mirrors, big white tips to black primaries, and broad white trailing edge continuing outwards as large white tongues/spots between the grey and the black – from below, extensive white and little black except on outer feather. Rather like mew gull, if not a bit more extreme (Thayer's like) – on the water, closed wingtip not so obvious, but white bases to primaries and big white tips to middle feathers visible beyond tertials. Unlike mew gull, the bill looked slightly longer than normal, not shorter.

Not-so-common gulls

Glaucous and Iceland gulls remain particular favourites. Drawings of Iceland gulls from long ago...

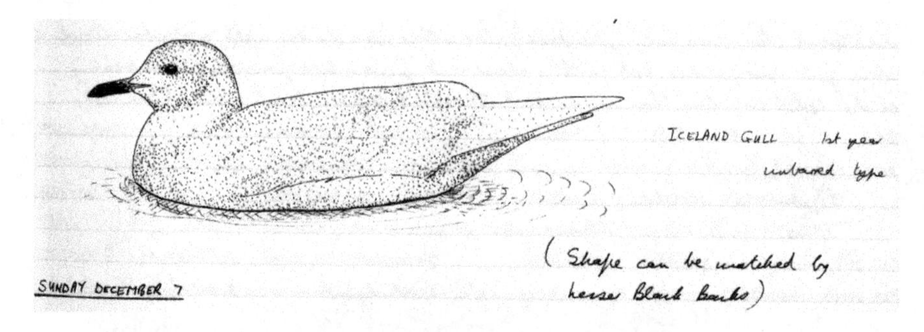

Iceland gull shapes are important in identification as well as always elegant and interesting. This is a pale, worn, 'unbarred' juvenile/ first-winter type (unlike most gulls, the juvenile does not moult in autumn, but just fades into a less-marked winter pattern).

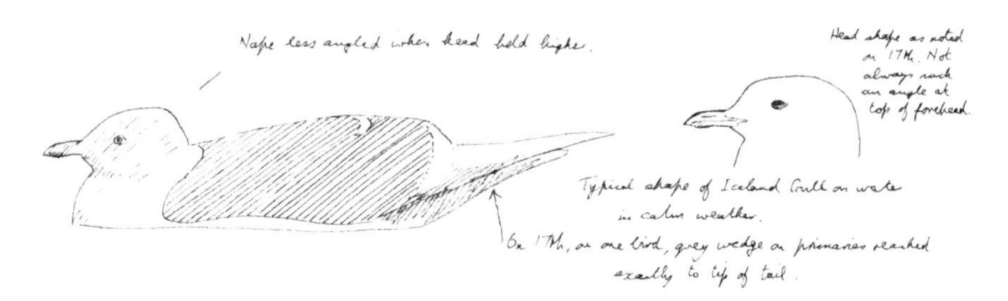

One of the adults with a marked 'tertial bulge' towards the rear: not all are like this.

How about two Iceland gulls off the beach near Bournemouth much more recently?

> I've seen Iceland gulls dozens of times, but never quite like this one. It was just off the beach and at first it was soaring high overhead in bright sun against a blue sky – I was so amazed and pleased to see it that I didn't notice the big wave that soaked my feet and drenched my rucksack, telescope and all. But it was worth the cold wet feet for an hour of brilliant Iceland gull, as it foraged, head to wind, over the surf, curled up and over the beach occasionally, dived into the breaking waves and later, very nicely, sat on the sea for a time to show the classic Iceland shape. It was lit by a strong sun plus the reflections from the waves and the white foaming surf, so the lighting effects were magical. At times there were one or two kittiwakes alongside and then a little gull appeared, itself bouncing about over the waves like a tern, its red legs dangling – fantastic! The kittiwakes often sailed along the face of the headland hill and cliffs.
>
> ...
>
> There have been two, if not three, Iceland gulls about recently, and there may have been one farther along the beach that I did not see today, but the overwhelming likelihood is that this was the same bird as the one I saw a week ago. It was in the same area, a little farther west along the beach below the cliffs. If it was different at all, it might have been a fraction smaller and a fraction more clearly marked, but I think this is more likely just a change in circumstances, light etc. It still looked a longish-billed bird. It was immaculate and beautifully marked and a classic Iceland in every respect. I have not enjoyed a bird so much for a long time, though. It was in a cold wind, with light from dull to bright sunlight, sometimes in stinging rain, but generally in good conditions. It was usually a few yards

offshore, riding the surf, in crashing waves, big rolling breakers and swathes of foam – it seemed almost certain at times to be overwhelmed, but always managed to stay perfectly happy and buoyant, only occasionally leaping up over a breaking wave and taking to the air, revelling in the conditions. Now and then, it did fly around a bit, and once came up over the beach and above my head. But it was best over the water – if you could leave aside the more distant surroundings, it was like something seen from an Arctic trawler! Brilliant. One kittiwake swept by so close that I thought it might hit me: several times it or another flew up to glide along the cliff face and away again.

...

Great stuff – good fun with the gulls, in sun, then heavy rain from deep purple-grey clouds, then sun, and showers of freezing, stinging rain and hail, with sudden strong squally winds chasing clouds of sand and spray along the beach.

I found an Iceland gull on the water just off the beach right away – I decided it must be the 'second' one reported recently, but it seemed a bit unlikely that one would replace the other doing the same thing, in the same place... But it did show some pale grey on the back and brown-barred greyish scapulars, and in bright light seemingly white on the forehead, throat and foreneck (later, this was not apparent). Also, it seemed to be more coarsely marked on the tertials/inner greater coverts and perhaps primary tips. The bill may have been different, but fine details were hard to see, both last time and this, on a bird on rolling waves (although much less spectacular today). Later, there were two birds side by, solving the problem!

One was clearly a first year, with more coarsely marked front and lower scapulars (rear scapulars finely patterned); darker around/behind the eye; darker on the tail with a more solid dusky band; almost unmarked on the primary tips. It had a dark eye, and the bill had the usual long dark tip with an S-shaped edge against a darkish yellow-orange-brown base.

The other bird had pale grey on the back, pale grey scapulars with fewer and weaker brown bars, but much more strongly barred tertials/inner greater coverts, a more uniform tail, more uniform wings in flight (less dark carpal mark perhaps) and seemingly a darker belly. The eye looked dark – I could not see any pale but it was hard to judge – but the bill might have had a tiny pale tip, then a slightly less wide blackish tip beyond a brownish area, fading to pale greenish at the very base. I think the bill of the first-year may have been longer, as well as a longer snout effect maybe – all subtle and hard to judge.

Anyway, all in all, some superb very close views – closer and better than before, if anything, and longer. The two were sometimes within a few feet,

Two superbly elegant and beautiful adult Mediterranean gulls, one red-legged (bottom), the other black-legged, in winter, back in the early 1970s. I used to make endless sketches of Mediterranean gulls: the species remains one that always gives me a bit of a 'kick' and is full of character.

> more often 30–50 yards apart and sometimes one or other would fly off out of sight along the beach.

Or Mediterranean gulls. These two were very early in the appearance and increase of the species on the Welsh coast, and I also saw the first for Staffordshire, but years later I was able to watch flocks several hundred strong in Hampshire. I've seen them over a local chip shop, over the garden. They routinely pick up crumbs amongst the crowds on Bournemouth beach and settle a few feet away. I once came across two young people photographing gulls as they dived around their heads and pointed out a Mediterranean, saying 'That's the one you need to get, it's pretty rare'... they looked at me as if I was really weird and quickly walked away without a word.

Perhaps I am. Probably.

Oh dear, this is becoming far too cluttered with stuff about gulls. Many times I had worked hard to find a yellow-legged gull or two amongst roosting herring and lesser black-backed gulls (although much later I was able to see them from a matter of two or three feet in such places as a grandstand overlooking the track at a Monaco F1 Grand Prix, a beach on Capri, the magnificent St Mark's Square in Venice and on my hotel room balcony in Portugal). My notes for a day in October 1995 – was it so long ago? – include the line 'This makes struggling to see a yellow-legged gull at a reservoir roost seem a bit silly.'

Martin Garner, introduced to me by artist David Quinn, was very much instrumental in sorting these things out and working out how to identify Caspian gulls, which have since become relatively widespread but still rare in the UK. I was taken to Mucking Tip in Essex by Martin, to spend a few hours watching gulls close to the Thames. There were between 150 and 200 yellow-legged gulls... and two adult gulls that we called '*cachinnans*?', which produced lengthy descriptions in our notes. We had Bob Glover, an Essex bird photographer, come along to record the birds, too. It was the first and only time I ever likened the gape of a bird – noted as vivid orange inside the corners – to the smile of Ramses II, the ancient Egyptian pharaoh. Pretentious? *Moi*? My scribbles include 'not yet on the British list', but Martin worked hard to get it there. It was later that he developed his 'frontiers of bird identification' website and deserved reputation, before his tragic early death from cancer.

***

Going back to some ideas discussed early in the book, it is vital to be sure you are aware of what you are really seeing, yourself, instead of what other people, or books, say you should be seeing. I well remember a single black tern amongst

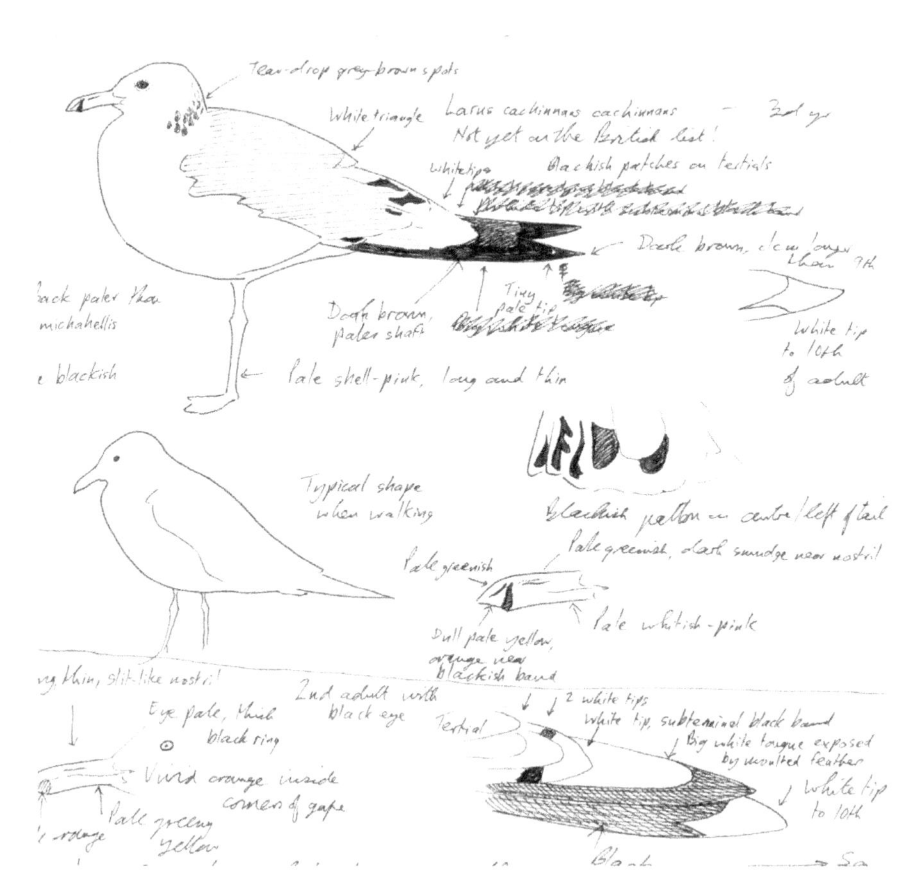

*I went to Mucking Tip again in August 1996, and the drawings above were one result. There were other drawings of various shapes, bill patterns, tail patterns of sub-adults and wingtip patterns, of what were then confidently entered as 'Steppe' gulls (*Larus cachinnans cachinnans*), now normally called Caspian gulls.*

common terns and black-headed gulls, being watched by several people in a hide. Look at that, some said, a classic black tern, look at the way it dips to the surface to feed, really distinctive, you can tell it right away. Well, it did dip to the surface to feed, that was not the problem, but in the particular conditions that day, over a freshwater lake, the common terns and even the gulls were dipping in *precisely* the same way. The black tern was identifiable by entirely different features, but not that one.

There was also much discussion about an American Bonaparte's gull, a brilliant find by Blashford Lakes warden Bob Chapman. A few people (some experienced and rarity-minded) talked through its features as they watched it, and noted how it 'differed' from the nearby black-headed gulls in various ways. Some of which I simply could not see.

> Better views than yesterday but still distant, often in difficult light with silvery water making it hard to see well. Frequently close to first-summer black-headed gulls – not a lot of difference at a distance! Clearly a little smaller, more delicate, slimmer necked, more tapered at rear; fine (black) bill not easy to appreciate much of the time. Regularly upended and turned frequently on the water, but black-headeds could do the same; also very quick and twisting in flight, but again some black-headeds did just the same!... In flight, odd upperwing pattern with a kind of middle stripe of white towards the tip, edged blackish; dark covert bar; underwing flashed bright white with black trailing edge on outer part – very distinct 'arctic tern' type wing, semi-translucent effect from both above and below. Even in a quick upward wing flick the white beneath was striking. In flight, wings narrow and tapered to finely pointed tip, more slender than black-headed. A good bird, but closer views would be appreciated!
>
> ...
>
> Bonaparte's gull – watched for a couple of hours – distant but better in clear light. On the water nearly all the time, only short flights – it didn't so much as raise a wing for half an hour or more at first! Compared directly with similar-age black-headed gulls – remarkably similar on the water, just a little smaller and 'greyer'. Although small and dainty, and said to upend and tip forward more to feed, it acted pretty much exactly like all the others, sometimes a little more active. The only apparent plumage differences at rest were: incomplete mottled hood a little less brown, more grey/black, extending a fraction lower on the nape; hindneck pale grey, less white – the latter two points only appreciable in a rear view. The tertial patch was darker than on most black-headeds, less faded brown (but not much different from others) – in a head-on view, as it tilted forward, it showed as a wider, more solid patch... The black bill was visible – but colour on black-headeds' bills was not always possible to see! But the bill was clearly thinner, but quite long. Pink legs easily seen as it took flight a couple of times. In the air, everything was different with the 'translucent' white underwing and black trailing edge instantly striking; wings looked slightly 'short' and tapered to narrow tips; tailband may be complete, but perhaps a small white break in the centre from beneath.
>
> ...

Mediterranean gulls – breeding and non-breeding – always exciting.

After a long time seeing little on the water, a 'pink' adult little gull flew left from nowhere, leading me to a group of gulls on the water right in the southwest corner, that I had overlooked – including the Bonaparte's gull. It very soon flew up, showing the instantly obvious underwing, but after that refused to even raise a wing – looking remarkably difficult in dull light, with some nearly identical black-headeds at that range. However, the blackish hood has firmed up a bit and now looks clearly a little bit extended over the nape, and also smudgily down the cheeks. The fine black bill is discernible but not too easy, and the size, intermediate between the little gull (on the water alongside) and black-headed, is a clue but not really obvious – a tricky bird! All the gulls were dipping forward, half upending, and spinning about, so the behaviour is not helpful.

...

Bonaparte's gull feeding on the water with a flock of black-headed gulls. It would be easy 99 times out of 100 to pass over it – knowing it is there, I look and find it within a minute or two, but without knowing, it would be overlooked unless it happened to lift a wing (which it does rarely!). Black-headed gulls alongside look practically identical, even to the hood tipping down over the nape a little (but perhaps less sooty blackish), although other first-summers look 'browner' on the dark parts and show obvious red/orange on the bill. It is certainly a touch smaller, especially head-on – but in the melee that is not always obvious (sometimes it just sits still a moment and then does look small and dainty). It is very easy accidentally to 'transfer' from the Bonaparte's to a black-headed for a few moments as they cross and re-cross in the ripples. As it tilts over, from behind even the closed wingtips seem to show a streak or two of translucent white surrounded by black. Once the wing is opened the underwing is striking, and today I saw it well several times, as well as the pink legs.

It is always important to be as objective as possible and not to get carried away with identification features that may be real enough, but might be very difficult to appreciate in reality.

Some more good gulls

I have found quite a number of rare or scarce gulls over the years, maybe because I spend too much time looking at them. But one day in October 2014 stands out. Fortunately, although it disappeared on the day, the bird in question was relocated by other observers a week or so later and remained to be seen and enjoyed by a great many people as it returned each evening to the roost.

It was sunny and mild, nearly calm, autumn colours starting to show – lovely day – and from the Tern Hide overlooking Ibsley Water, with binoculars I saw a particularly white flashing underwing across the lake, over the trees along the A338 or beyond them into the Avon valley (miles away!). Despite the white, Mediterranean gull never entered my head, as there seemed to be a dark wingtip but nothing like second-year Mediterranean streaks – more like a kittiwake, in a very odd place. I still thought kittiwake when I mistakenly tried to get it in the telescope instead of looking harder – but failed, saw it in binoculars again and tried once more, failing again to pick it up as it twisted against the sky. I thought I had lost it, wondered vaguely about Franklin's gull, a vagrant from North America, but with not much hope – but it reappeared, I did get it in the scope, and there it was! A Franklin's gull, for sure.

It was flying quite fast, with just two or three black-headed gulls nearby usually, with wings markedly angled (both back and down) and quite quick shallow beats, with striking sideways twists and turns, frequently doubling back and rising higher, then swooping down again. I had a short look at it, went off to try to tell people at the visitor centre, then looked briefly from the main roads, before returning to the hide and finding it again. This was between about 12.00 and 12.30. For the next 20–30 minutes I saw it several times more, helping other people to see it but probably none of them to identify it as such, and had excellent views of it in full sun against darker clouds. It seemed to head off farther south eventually.

This would be at about half a mile range, sometimes a little more, coming a little less than that at best, viewed with ×10 binoculars and ×30 telescope. Light excellent throughout, telescope supported on hide ledge.

Black-headed gulls alongside showed the wing pattern above and below perfectly well. The Franklin's showed a gleaming white underwing with small 'dipped in' black tips, and the upperwing was steely grey, darker than a common gull or kittiwake I should think, paler than a *graellsii* lesser black-back but not much, blending (?) back into a very broad and really striking white trailing edge. The white curved between the grey and small black wingtips – there was no fading to a paler grey as on kittiwake – and I was quite quickly able to be absolutely sure that there were small white tips at the extreme end of the outer feathers, beyond the black. The white across the primaries inside the black later seemed to resolve into spots or rounded patches at closer range, rather than a simple band, but clearly enough of a band, or enough white, to make it a definite adult, not a second-year type.

To finish with gulls, for now, two more really good ones: a Ross's gull in March 1995, near Liverpool...

The back was steely grey, and usually the round, dumpy-looking head gleamed contrastingly white, with a faint smudge around and over the eye – likewise, closer, this resolved itself into a small, curved hood between white forehead and nape. Bill seemed stubby and dark but no detail seen. No eye-ring detail visible at this range. The tail looked white but several times, with surprising clarity, I saw a slightly streaked area of grey at the base – seemingly from the base of the tail on the middle two-thirds or so, fading out towards the tip. I was a little surprised how easy this was to see.

The wingtip pattern might have made it a little illusory, but it looked shorter- and rounder-winged than a kittiwake or black-headed gull, a little broader in the wingtip.

...and a Sabine's gull in the unusual month of July, on the Norfolk coast.

On reflection, the amount of black in the wingtip and the apparent distinct white primary tip spots suggest a second-winter, but at this distance, with the detail observed, it is unsafe to say more than second-winter/adult.

I saw it again a few times, then one afternoon tried to get closer but failed:

> I gambled and went to the Lapwing Hide, looking into the setting sun: I hoped that as the sun went down the light would improve but in fact the water remained bright and the gulls, spread all over the lake in a huge extended flock, just remained silhouettes. Earlier, the light varied from fantastic to impossible for the gulls – had the Franklin's come in early it might have been great! But I don't think it was seen at all today.

> Having said that, having watched many hundreds of gull roosts over the years, few have been so good as this just to enjoy looking at – a beautiful sky with great cloudscapes, a fine sunset, and thousands of bright blue gulls on shimmering gold and orange water – just sensational.

A couple of months later it was gone, but was replaced as the centre of attention by a ring-billed gull – my second American gull find in a few weeks. For several winters, a ring-billed became a regular bird there, and on some days there were two or three, once conceivably four. It was always a good, fun challenge to try to find one in the flocks of thousands on the water.

Gulls, eh... always worth a look.

SEABIRDS AND SUNSHINE

Skomer

There can hardly be a seabird colony that isn't beautiful, if only by virtue of its birds, the wash of the waves, the scent of sea and fish. But some are better than others, no doubt. The brilliant cliffs of Duncansby Head at the northeastern corner of Caithness make up one such place. The great red cliffs and pointed stacks are beyond my powers of description, certainly. The last time I went, there was a thick fog. Still, there's always hope of another.

June:

> I have seen more spectacular seabird colonies, but the beautiful island of Skomer proved to be quite something. All the seabirds were seen for a long time during seven and a half hours on the island, plus half an hour in the boat.
>
> The auks were all superb: razorbills and guillemots on the ledges and out to sea; razorbills frequently showed their bright orange-yellow gapes. They seemed to be incredibly smooth and velvety, with a soft bronzy sheen on their black necks. The guillemots looked, by contrast, quite pale, a soft grey-brown. Despite much close examination I saw only one bridled bird with its white 'spectacle' line. The puffins were particularly good, as puffins always are. Many were on slopes that were riddled with burrows, and watched displaying and arguing on rocks and at the entrances to their holes, but most were on the water and, in the mid-afternoon particularly, large rafts formed on the sea, close inshore. These were remarkably tame so we could watch puffins by the dozen, in bright sunshine, idly swimming barely 30 yards away. Amongst these were a few of the larger auks and all three could be watched as long as you liked at close range. From the boat you could look at them on their ledges just seven or eight yards off. In the South Haven, between High Cliff and The Neck, a very large number of puffins gathered; later there was another large flock in North

> Haven where I counted 730, but the earlier group was larger. Several were at The Wick, more between Bull Hole and Garland Stone. Occasionally 200–300 would fly up from the sea to visit the burrows. Altogether, some of the very finest views of these three auks that I have ever had, in some of the most beautiful settings. Kittiwakes were especially good at High Cliff and The Wick; I managed to see several eggs, while the calls of kittiwakes rang around the cliffs together with the superb whirring of auks and cackling of fulmars. The sight and sound of such a mixed colony is unbeatable. Lesser black-backed gulls had several sub-colonies in a beautiful sea of bluebells and red campion. Choughs were in view, briefly, from the boat as we arrived, but not seen again until we were waiting for the boat to leave, when they appeared over The Neck.
>
> Almost as soon as we landed a short-eared owl appeared by the warden's house and eventually settled deep inside the bluebell carpet; later marvellous views of it in the air and perched – several views at various places around the island, often chasing gulls and crows, or sitting out on open grass, giving unbeatable views.
>
> The whole island is full of holes. We found some remains, but never saw a Manx shearwater even though 50,000 pairs breed here! They are visible only at night. But there are some 7,000 pairs of puffins, 500 pairs of storm petrels (also nocturnal and not seen) but only two pairs of choughs.
>
> From the western end of the island I looked out to sea and could make out the small, domed island of Grassholm, with a mass of birds in the air above it and a broad white band across one side: part of the great gannet colony there.

My notes gave rough totals of 20 shags, 200 razorbills, 500 guillemots, 2,000 puffins, 30 fulmars, 2,000 kittiwakes, 500 lesser black-backed gulls, some stock doves, buzzards, two choughs, a raven, two short-eared owls. I expected to find larger seabird numbers, but the island was, and is, intoxicating. It has, needless to say, been a great favourite of a number of artists and photographers over the years, as well as scientists researching seabird biology.

One point that needs to be made somewhere: seabirds used to take an enormous hit from pollution, but now take a major hit from lack of food in the seas. Years ago, I walked regular routes along beaches to count dead birds on the tideline, and to record any degree of oiling (in the Beached Bird Survey). I was lucky and found rather few, but there were incidents, at times, all over Europe, in which sometimes tens of thousands of birds (principally auks) were killed by oil spills.

These are perhaps a little bit forgotten now, but, as with so many things, younger readers ought to be aware that such things were frequent not so long ago.

While I was writing this, the awful reality of avian flu began to hit television news bulletins each evening: a sure sign that something is really wrong. Scores of dead great skuas, hundreds, perhaps thousands, of dead gannets, and dead and dying auks by the thousand... an appalling state of affairs. Let's hope things will have improved by the time you read this, but in 2023 it was returning with a vengeance.

A good place to see a shrike: Grassholm

It's 15 September, a bit late in the year. In a box I have a Manx shearwater, picked up from Llandrindod Wells. It is a wind-blown bird from the Irish Sea, and in the box even gives a blast of Manx shearwater crowing, cackling calls.

In Pembrokeshire, I release it over the sea and, lively as ever, it flies off strongly. Good deed done. With Pembrokeshire island expert David Saunders (who organised the brilliant 1969 Operation Seafarer seabird census and wrote up the results in a fine book), I get into a little boat and head off west from Dale. We stop off at Skokholm, famous for Ronald Lockley and his studies of Manx shearwaters and puffins. I remember Carol and Caroline in his book, shearwaters nesting on a hummock behind his house. It was my first visit, but lasted barely an hour, hardly enough to appreciate the island.

Skomer, not far away, is another beautiful place and I had been there on much longer visits, but have never stayed overnight for the nocturnal shearwater arrivals. After a summer visit, you come away smelling of bluebells; colonies of lesser black-backed gulls look amazing, blobs of immaculate white and slate-grey picked out with spots of vivid yellow, in a sea of lilac-purple-blue bluebells and red campions.

It was impossible, anyway, to do justice to the next island, even with a three-hour stop-off. The atmosphere and character of Grassholm, with its constant noise, the smell, the air and sea and ground all full of gannets, was tremendous. It is just an offshore rock, really, 11 miles from the mainland.

There were plenty of gulls, mostly on the south side of the rock, dominated by 800 or more great black-backs, a remarkable number for a species usually in much smaller numbers than others on a beach. Only herring gulls scavenged around the gannet colony fringe. Kittiwakes were just hanging around, adding a bit of beautiful noise to the general confusion of semi-mechanical sounds.

The oddest bird, though, in a rocky gully just above rolling waves, was a juvenile red-backed shrike! The whole place is pretty much devoid of vegetation, or at least anything for a shrike to hide in, and it was odd to see this bird, a migrant from continental Europe, in such a bleak setting, so far west. It was hunting insects, as best it could, and I looked down on it from higher rocks, so had a slightly odd view of a broad-bodied bird with drooped wings, revealing the roundness of its body.

There were other migrants, too: meadow pipit, yellow wagtail, pied wagtails, wheatear, stonechat, redstarts, swallows, willow warbler, chiffchaffs, garden warbler, several goldcrests: quite a decent haul. I supposed rock pipits were regulars all year.

Yet these were all playing second fiddle to the main event. Grassholm has one of Europe's biggest gannet colonies: at the time some 30,000 pairs. There were beautifully crisp black-and-white aerial photographs on the walls of the RSPB Wales office in Newtown, from which accurate counts could be made from years gone by. It proved to be a wonderful, tumultuous, amazing sight. A constant clamour created a curiously mechanical, clattering effect (seeming to merge into a synchronised, rhythmic *uck-erruck-erruck-erruck*), with individual deep *urrah-urrah* calls and odd whistling notes from the juveniles. If we, by accident, approached a bit too closely or too suddenly, the edge of the colony would erupt into noisy, thrashing confusion, birds all tumbling about: one group, evidently non-breeders at the fringe, cleared off, half-scrambling, half-flying, throwing up a cloud of dust, feathers and guano.

But a careful, quiet approach allowed me to photograph birds from as little as eight feet (my longer lens focused down to 8 feet 6 inches, and, careful as I was, most photos proved to be just a bit off). Adult gannets would sail in to land within a few feet of me, taking no notice whatever, with all kinds of bill scissoring, fighting, weed-tossing and calling going on as if I was not there. A humbling experience. There were some flying young, and several on the sea, but many were still in the colony at nests and some were still thickly clad in down on the head and neck, showing a little less synchrony than expected at a late date. There were several thousand birds in a long string on the sea, going a mile or more to the north, and maybe 5,000 in flight, yet the numbers on the ground seemed undiminished.

A gully held several dead, some injured and some trapped birds: one hung by a leg on a long piece of nylon twine and there was fishing netting, ropes and bits of polythene everywhere. It was a sensational, stunning place yet marred by so much dangerous mess. Thousands of nests were built of weed and fishing netting or twine in equal amounts. Instead of a seabird city, we had come to a bird slum. Roger Lovegrove sometimes organised late-autumn clean-up visits, an effort to reduce the annual misery and mortality of gannets old and young caught up in non-degradable rope or mesh.

The colony then had hardly spilled over the top of the island, so it was possible to walk to the central ridge and look over, across the great mass of gannetry. It has since expanded, creeping over the top, making the ridge inaccessible and the views from the land a little less comprehensive. Old birds would fight over their tiny territories, and youngsters stepping out of line were pecked around the head: one I picked up, to release it from a mass of netting, was smeared with blood. It was a phenomenal place: not all sweetness and light, but astonishing.

Gannets, gannets, gannets: thoroughly extraordinary birds, whether battling with their neighbours or sailing so perfectly, so beautifully, against a blue sky.

The effect was outstanding: surely one of the world's great bird sights. It fulfilled an ambition I had had for years. I wrote in my notes, 'Now I just want to get back.'

I never did, although I did fly over the island, circling in a little light plane at a safe height: for just a little while becoming a passing gannet floating in the clean, crisp Irish Sea air.

Incidentally, the gannet, and its relatives worldwide, have been subject to intense research by many people and an early monograph – a study of a single species – was written by J. H. Gurney (and was being revised or rewritten by James Fisher, but not completed before his fatal car crash). But for me the real gannet man was Bryan Nelson, who not only studied them but wrote about them in an inspiring and evocative way. If you can, beg, borrow or buy one of Nelson's gannet books: they are classics of their kind. The illustrations by John Busby are brilliant, too, showing how it should really be done.

Some years since I had been to such a decent seabird colony I travelled to South Stack in North Wales to write an article for *Birds* magazine. I wrote that, coming back to a fine seabird cliff, I brushed a tear from my eye: just the wind... of course, it wasn't. Such places have a strong emotional effect. South Stack is brilliant: I had not, perhaps, given it sufficient credit in my mind, forgetting how good it really was until this return visit. Peregrines gave close views. This is the area where great bird artist Charles Tunnicliffe, who lived at the other end of Anglesey, watched and painted them around their nesting ledges. Bempton

A Grassholm gannet with its full-grown chick almost ready to fly. It will take several years to become fully adult in plumage, and the intermediate, more or less chequered, stages are not always easy to fit precisely into year-groups. Unusually, sometimes an individual will have different patterns on each wing, rather than the symmetrical mirror-image that is the norm.

Cliffs in Yorkshire are of a different nature, sheer limestone, but with an abundance of nesting seabirds; Gower has much more irregular cliffs with grassy slopes and upstanding rocky headlands but, Worm's Head apart, few seabirds. Pembrokeshire has a good deal of red sandstone; the east coast of Scotland has a succession of strong sandstone cliffs. In the end, perhaps, sandstone wins out. Or the sparkling limestone in summer sun against a blue sea. No, the sandstone with all its narrow horizontal edges and dark cavities.

The tortured, twisted metamorphic rocks at South Stack on Anglesey offer plenty of opportunities for busy, noisy colonies of guillemots (here with a couple of razorbills) and kittiwakes. Unlike many such colonies, they are also enlivened by choughs.

Going to sea

For many people the sea holds a strangely irresistible pull, even if there is no history of seafaring in the family. Read the first pages of *Moby Dick*... We go to the sea because it is wild, untouched, untamed, open and free and wonderful. It dominates us, rather than the other way round. There is little that we can do to change it, to bend it to our way of thinking. Which is why I dislike the wind turbines that urbanise it, that destroy the wildness and romance, however much we need them.

Swansea University had a small marine research vessel, the *Ocean Crest*; while Mumbles Pier was a stop-off for steamer trips around the Bristol Channel. Money was tight so the latter was a very rare treat, and only once, in about 1970, did I get onto the *Ocean Crest*, essentially a vessel used by a different department. This happened to be just two days before the steamer trip, in June. The *Ocean Crest* left Swansea docks and spent a while dredging in Oxwich Bay, before two long 'trawls' off the western end of Gower and into Carmarthen Bay.

Seabirds on Gower were mostly limited to a small kittiwake colony on the mainland, plus a mixed colony on Worm's Head, onto which I occasionally scrambled at low tide to enjoy the gulls and auks, and to sit by the blowhole hoping to get caught in the roaring jets of spray. Off headlands farther east, such as Mumbles, the variety offshore was usually limited and numbers very small, even when the sea was in turmoil. Sometimes far out there was a 'super wave', a huge pile of water way above the general level, topped by white foam, moving rapidly up channel.

It was interesting, then, to see from the boat a scattering of Manx shearwaters including a flock of 55, three razorbills, a few guillemots and three gannets. This trip seemed to have possibilities. Off Oxwich, a fabulous headland, bay and marshy area mid-Gower, shearwaters totalled 86 and a fulmar added to the variety. In Carmarthen Bay we saw more auks, a scattering of shearwaters and terns (some undoubtedly arctic) and then a surprise, for the mid-June date, a great skua. By then there was quite a knot of birds following us, including fulmars, kittiwakes and a selection of commoner gulls, and the skua was drawn in to chase these in turn: excellent and unexpected.

The return trip was similar but with shearwater numbers reaching almost 200. This would have been unusual from land on Gower, although Manx shearwaters breed in huge numbers not so far away in Pembrokeshire, on its beautiful scatter of islands – Skokholm, Skomer, Ramsey.

If you have not seen a shearwater, then be aware that it lives up to its name... a Manx flies with rather straight, stiff, if sometimes slightly arched, wings, tilting over sideways to present the underside of the wing to the wind. It is not a sailplane so much as a yacht, using its sails to progress across and against the wind. In a good wind, the shearwater hardly beats its wings; in calm weather, it has a short series of quick, flickering beats between short glides, but not so quick as, say, a razorbill or puffin. Black above and white below, it looks first like a black cross, then a white one, as it tips over and back again. My first experience was in western Scotland, where flocks gathered ready to go ashore each evening on Rum. They were very distant, minute but very special, and I remember the trusty Fitter and Richardson guide had them just right, like little black crosses low over the distant waves.

At Skomer and Skokholm, many thousands nest in burrows, in which they remain hidden by day. Incoming birds arrive after dark, announcing their presence with a loud, rhythmic crowing (*it-i-korka!* is a way to write it down, but that doesn't do it justice at all). To my shame, I have never been to a Manx shearwater colony after dark, a gapingly large hole in my experience.

I did, though, visit La Gomera in the Canary Islands and found, quite unexpectedly, that the hotel was on top of a cliff with a colony of Cory's shearwaters. Cory's are big, powerful birds compared with Manxies. It was a fantastic experience to have Cory's whistling within feet of my head, landing just a few feet away on a gravelly patch at the top of the cliff where it was light enough to get detailed views of their beautiful expressions and curious hooked bills, and creating a wonderful chorus of the strangest metallic, buzzing and groaning noises. Often they could be heard calling out on the sea, where they rested until it was dark enough to come in, with an assortment of dolphins and small whales for company. It gave me an idea of what the Manxies might be like.

On boarding the tourist steamer at Mumbles Pier, it seemed that a few seabirds might come our way: the previous year we had done the same and managed to see some storm petrels between Ilfracombe and Lundy Island. These were, I have to say, the real target, something very special. This time, we left Mumbles and immediately found shearwaters and then, barely a mile offshore, the first storm petrel – a great start. From Mumbles to Ilfracombe, a few auks appeared, including guillemots, razorbills and a puffin, and four or five storm petrels, a result of constant scrutiny of the waves around the ship. The petrels followed us for half an hour, usually hanging back up to 200 yards astern, gradually

Don't necessarily expect the best views of small seabirds such as the storm petrel over open sea, from a ferry, still less a cruise ship: instead, enjoy the whole idea of this tiny bird surviving in its element, the vast open ocean, navigating to and from its nesting burrow while having nothing more than a very limited view from a low-level point above the waves.

drawing closer until perhaps 30 or 50 yards away and then swinging off right out of sight, before reappearing far back in the wake. In bright sun, they looked very brown, with the bold white rump very striking and a white underwing mark often visible. But on this occasion there were no more storm petrels on the Lundy section after Ilfracombe, nor on the return: indeed, no more all day.

Occasionally a storm petrel was seen from the Gower headlands, including Mumbles, not usually in a gale but during the summer when they are more often in the Bristol Channel. You might get lucky at a headland such as Portland Bill. On rare visits I have seen more from land in southern Ireland, where observations from the headlands seem just routine, but it is best as a rule to try to get out on a boat to see them. They are well worth it.

Most memorable, I once joined a group of ringers at night, catching storm petrels by playing their calls to lure them ashore – and having one of those tiny seabirds briefly sitting warm-footed on the palm of my hand before it flew off into the night to resume life at sea was humbling.

On the other hand, you can enjoy beautiful views of gulls such as this lesser black-back drifting along in the slipstream. This one came so close it reminded me of a cousin who farmed in Essex and said he could sometimes reach out and gently tug the wingtip of a passing gull as he sat on his tractor, ploughing a field.

Some years after my short cruise in Carmarthen Bay, I flew over it in a small Cessna several times, with Roger Lovegrove, counting common scoters from the air. The bay had (and presumably still has) a large population of scoters in late summer and autumn, and this was an innovative way to monitor their numbers in an area that is hard to see properly from distant headlands and low beaches.

Watching seabirds from a vessel such as a ferry or even a cruise ship can be great fun. It might be a boat to Ireland or across the Channel, or to the Western Isles or Northern Ireland, even a tourist trip. You need to choose your spot: try to see what is up ahead and heading your way, then scan slowly back towards where you have come from; then quickly scan back round again (don't 'jump' back to the front in case you miss something). At the stern you will see things in the wake but can't see up ahead too well, and the deck might be vibrating too much so you can hardly hold binoculars steady (or you get in the smoke from the funnel). At the front, you will be exposed to the wind as you steam forwards, but get a good view all round. Find a good, sheltered spot with as little noise and vibration as you can, but you might have to accept just a one-sided view of the passing sea, so you just get a bit more than a 50% chance of seeing the next yellow-nosed albatross.

FROM A DISTANCE

This book should by now have impressed on the reader, your good self, that birdwatching is what you make of it: the more you put in, the more you get out. But – although it is entirely up to you what you do with it – you may have little control over how you react to the birds you see and places you visit. For me, it is often very much an automatic emotional response, however strong or mild that might be. Such a thing is personal and by its very nature unique: you will not necessarily feel the way that I do about seeing a guillemot colony or a displaying black grouse, although you might be equally excited by a new Siberian vagrant.

In any case, it does perhaps seem too much like looking backwards, reliving old memories, and that should not always be the case. After all, something very good indeed may be waiting around the corner.

There is something extra-special about seeing birds really close up. We appreciate the confidence they show. And we all like good views of whatever we see: every feather, every scale on the feet as a chaffinch hops over a pub garden table, or a siskin hangs on the feeder by the window, or a dusky warbler creeps through a bramble thicket. But less has been said about some birds that have special qualities even from a distance.

Can you really get excited about distant dots? Well, yes, why not? I well remember that hot sunny day, already described, in Inverpolly, that magical expanse of low, rolling country in the Highlands, dotted with blue lochans, where great peaks – Cul Mor, Cul Beag, Canisp, Suilven, Stac Polly – rise up from nowhere. Suilven, scanned from some ancient Kodachrome, grandly announces itself as one of the desktop images on my computer screen. Golden eagles over one peak were five and a half miles distant. But they were eagles, and they were great. That's what eagles do: that's where they *belong*.

Now, after a day missing nearly all of the birds that others will later record on the website, at Hengistbury Head, I sit on the beach, scanning an improbably bright blue sea, from Durlston Head and Purbeck to The Needles. The hot sun burns my face while all the tarmac paths, fences, visitor centres, signposts, multicoloured beach huts, dog-waste bins and general suburbanisation of the once-wild nature reserve are all firmly behind me, excluded from my consciousness. There, over a usually sublime yet rather birdless sea, are gannets. Just two or three, a mile away or more, circling and now and then diving with that characteristic high, white, silent waterspout splash. Gannets: they take me back to the glorious west coast of the Scottish Highlands where I loved to watch them years ago, and I still remember so clearly my first one ever, sailing past Ardnamurchan Point with its elegant lighthouse and views around the isles. Yes, I like closer views, of course, but these distant ones – silent, long-range, far removed from me and everything to do with the land, living lives so precisely matched to their maritime habitat – they have a special something all of their own.

True, it is a mixture of nostalgia and the pull of the sea – the lonely sea and the sky and all of that – and maybe the chance to dream about distant Grassholm or Noss and other great gannet rocks. But an extra something gives the whole experience that bit more meaning, distant though the gannets are. Come closer, and the effect is memorable, but an altogether different one.

Ardnamurchan Point, and my first gannet.

Or you might be watching gannets from Bournemouth beach. Turn your back on the beach huts, car parks and 122 dogs (I've counted them) and even a distant gannet – perhaps especially a distant gannet, like the left-hand one – can take you off to wherever on the coast you want to be. And, however grey the sky, there will always be a ray of sunshine eventually.

Birdwatching can do that.

ACKNOWLEDGMENTS

Several people encouraged me to continue writing this book through a long gestation period. The changes they suggested helped turn a long aimless ramble onto a much more manageable and direct route. In particular, I must thank Andy Swash, Janice Hume and John Thorogood for their positive comments and support. Conor Jameson and Steve Holmes kindly read the text and made invaluable suggestions that helped me improve the final book. Nigel Massen at Pelagic Publishing quickly offered me the opportunity for publication, for which I am hugely grateful, and David Hawkins and Hugh Brazier then used their professional design and editorial talents to excellent effect and kept me optimistic about the outcome. As ever, however, any errors of fact, opinion or judgment remain my own, so please laugh at me, not them.

BIRDWATCHING RULES AND THE BIRDWATCHER'S CODE

Rules? Basically, there aren't any... enjoy it and do it your way. There are, however, some things to keep in mind. Many years ago, I was involved in drafting the first 'Birdwatcher's Code of Conduct' with ten or a dozen bullet points, something that perhaps gradually faded away over the years, but which still highlights some important issues.

The RSPB and BTO version now has just five points to remember:

1. Avoid disturbing birds and their habitats – the birds' interests should always come first.
2. Be an ambassador for birdwatching.
3. Know the law and the rules for visiting the countryside, and follow them.
4. Send your sightings to the County Bird Recorder and the Birdtrack website.
5. Think about the interests of wildlife and local people before passing on news of a rare bird, especially during the breeding season.

The **Scottish Ornithologists' Club** introduces the code like this:

- Please help everybody to enjoy birdwatching by following the code, leading by example and sensitively challenging the minority of birdwatchers who behave inappropriately.

The Birdwatchers' Code puts the interests of birds first and respects other people, whether or not they are interested in birds. It applies whenever you are watching birds in the UK or abroad.

The point about behaving properly when abroad is a good one: it is too easy to forget in new and exciting places.

The **BTO** expands the points in the Code:

- Whether you are particularly interested in photography, bird ringing, sound-recording or birdwatching, remember always to put the interests of the birds first.
- Avoid going too close to birds or disturbing their habitats – if a bird flies away or makes repeated alarm calls, you're too close. If it leaves, you won't get a good view of it anyway.
- Stay on roads and paths where they exist and avoid disturbing habitat used by birds.
- Think about your fieldcraft. You might disturb a bird even if you are not very close, e.g. a flock of wading birds on the foreshore can be disturbed from a mile away if you stand on the sea wall.
- Repeatedly playing a recording of bird song or calls to encourage a bird to respond can divert a territorial bird from other important duties, such as feeding its young. Never use playback to attract a species during its breeding season.
- Respond positively to questions from interested passers-by. They may not be birdwatchers yet, but a good view of a bird or a helpful answer may ignite a spark of interest. Your enthusiasm could start a lifetime's interest in birds and a greater appreciation of wildlife and its conservation.
- Consider using local services, such as pubs, restaurants, petrol stations, and public transport. Raising awareness of the benefits to local communities of trade from visiting birdwatchers may, ultimately, help the birds themselves.
- Follow the codes on access and the countryside for the place you're walking in.
- Irresponsible behaviour may cause a land manager to deny access to others (e.g. for important bird survey work). It may also disturb the bird or give birdwatching bad coverage in the media.

Legislation provides access for walkers to open country in Britain, and includes measures to protect wildlife. Note that the rules and codes are different in each part of Britain, so plan ahead and make sure you know what you can do.

- In **England**, the Countryside Code and maps showing areas for public access are online at www.countrysideaccess.gov.uk.

- In **Wales**, access maps can be found at www.ccw.gov.uk/tirgofal, and the Countryside Code at www.codcefngwlad.org.uk.
- In **Scotland**, access is available to open country and to field margins of enclosed land to reach open country, provided you act in accordance with the Scottish Access Code – see www.outdooraccess-scotland.com.
- Although there is no statutory right of access in **Northern Ireland**, there is lots of information, including the Country Code, at www.countrysiderecreation.com.

If you witness anyone who you suspect may be illegally disturbing or destroying wildlife or habitat, phone the police immediately (ideally, with a six-figure map reference) and report it to the RSPB.

Remember, there are plenty of other things to watch if the birds are a bit quiet – like this chub and banded demoiselle.

INDEX

Printed in the USA
CPSIA information can be obtained
at www.ICGtesting.com
JSHW010811210124
55624JS00001B/1